CLIMATE CHANGE ADAPTATION

COLUMBIA UNIVERSITY EARTH INSTITUTE
SUSTAINABILITY PRIMERS

COLUMBIA UNIVERSITY EARTH INSTITUTE
SUSTAINABILITY PRIMERS

The Earth Institute (EI) at Columbia University is dedicated to innovative research and education to support the emerging field of sustainability. The Columbia University Earth Institute Sustainability Primers series, published in collaboration with Columbia University Press, offers short, solutions-oriented texts for teachers and professionals that open up enlightened conversations and inform important policy debates about how to use natural science, social science, resource management, and economics to solve some of our planet's most pressing concerns, from climate change to food security. The EI Primers are brief and provocative, intended to inform and inspire a new, more sustainable generation.

Renewable Energy: A Primer for the Twenty-First Century, Bruce Usher

Climate Change Science: A Primer for Sustainable Development,
John C. Mutter

Sustainable Food Production: An Earth Institute Sustainability Primer,
Shahid Naeem, Suzanne Lipton, and Tiff van Huysen

*Managing Environmental Conflict: An Earth Institute
Sustainability Primer*, Joshua D. Fisher

CLIMATE CHANGE ADAPTATION

AN EARTH INSTITUTE SUSTAINABILITY PRIMER

LISA DALE

Columbia University Press   *New York*

Columbia University Press
Publishers Since 1893
New York   Chichester, West Sussex
cup.columbia.edu

Library of Congress Cataloging-in-Publication Data
Names: Dale, Lisa, author.
Title: Climate change adaptation : an earth institute sustainability primer / Lisa Dale.
Description: New York : Columbia University Press, 2022. |
Includes index.
Identifiers: LCCN 2021059530 | ISBN 9780231199162 (cloth) |
ISBN 9780231199179 (paperback) |
ISBN 9780231552974 (ebook)
Subjects: LCSH: Climatic changes. | Climate change mitigation. |
Global environmental change.
Classification: LCC QC903 .D25 2022 | DDC 363.738/746—dc23/
eng/20220128
LC record available at https://lccn.loc.gov/2021059530

Columbia University Press books are printed on permanent
and durable acid-free paper.

Printed and bound by CPI Group (UK) Ltd, Croydon, CR0 4YY

Cover image: Balate Dorin / Shutterstock.com

CONTENTS

LIST OF ACRONYMS

AF—Adaptation Fund
BGI—blue-green infrastructure
CBA—cost-benefit analysis
CDP—Carbon Disclosure Project
COP—Conference of the Parties
CRM—climate risk management
CSA—climate-smart agriculture
CSO—civil society organization
DALY—disability-adjusted life year
DRR—disaster risk reduction
ECA—Economics of Climate Adaptation
EU—European Union
EWS—early warning system
FbF—forecast-based financing
FCIP—Federal Crop Insurance Program
FDI—foreign direct investment
FEMA—Federal Emergency Management Agency
FFS—farmer field school
GDP—gross domestic product
GEF—Global Environment Facility
GHG—greenhouse gas

IGO—intergovernmental organization
INDC—Intended Nationally Determined Contribution
IPCC—Intergovernmental Panel on Climate Change
IUCN—International Union for the Conservation of Nature
LDC—less developed country
NAPA—National Adaptation Program of Action
NAP—National Adaptation Plan
NbS—nature-based solutions
NDC—Nationally Determined Contribution
NFIP—National Flood Insurance Program
NGO—Nongovernmental organization
NIMBY—Not in My Backyard
OECD—Organization for Economic Cooperation and
    Development
PA—Paris (Climate Change) Agreement
REAP—Risk-Informed Early Action Partnership
ROI—return on investment
SDG—Sustainable Development Goals
SIDS—Small Island Developing States
UN—United Nations
UNDP—United Nations Development Program
UNDRR—United Nations Office for Disaster Risk Reduction
UNFCCC—United Nations Framework Convention on
    Climate Change
UNISDR—United Nations International Strategy for Disaster
    Risk Reduction
UK—United Kingdom
VRA—Vulnerability and Risk Assessment
WFP—World Food Program

CLIMATE CHANGE ADAPTATION

INTRODUCTION

CLIMATE CHANGE: MITIGATION AND ADAPTATION

Scientists first identified the connections between the industrial release of carbon dioxide and the stability of the planet's temperature in the late nineteenth century. It took nearly a century for these early scientific observations to become matters of broad social and political concern, as growing certainty about a warming world entered the mainstream consciousness only in the late twentieth century. Most of the attention was devoted to one central problem: how to reduce these harmful emissions within the context of continued economic growth and human development. Formulation of climate mitigation policy began in earnest with the 1992 establishment of the United Nations Framework Convention on Climate Change (UNFCCC), and the result has grown to encompass all nations on earth. But until quite recently, the notion of adapting to a changing climate thoughtfully, systematically, and with advance planning was not considered a global priority.

Part of the reason the global policy machine has been slow to incorporate adaptation is that many initially saw the task as a tacit admission of failure. Mitigating climate change seemed both desirable and possible, and in the early years of policy development, the focus was almost exclusively on national efforts to reduce emissions while establishing tracking mechanisms and holding polluters accountable in global governance settings. But now it is clear that most countries have failed to implement sufficiently aggressive mitigation policies, and since climate change is by definition a global problem that defies borders, even countries that have realized the most robust greenhouse gas (GHG) emissions reductions haven't halted global climate change. Temperatures have already increased, seas have risen, precipitation patterns have shifted, and impacts are being experienced across much of the world. Adaptation is therefore already happening. Farmers are choosing different crops as they see the seasons shift. Coastal developers are reconsidering big infrastructure projects as rising seas threaten existing structures. Humans are migrating and settling—in many cases without full awareness of the driving forces—in places that offer better climactic conditions.

Scientists also now know that mitigation and adaptation are not zero-sum. The two policy streams can be integrated, and even isolated actions can be mutually reinforcing. Further, adjacent global policy efforts that seek to address improvements in health outcomes, economic growth, educational attainment, environmental protection, and other dimensions of sustainable development are frequently aligned with best practices in adaptation. These data points confirm that the challenge of adapting to climate change is not a siloed endeavor. It isn't even new. It is rooted in decades of international peace building and the advancement of human societies.

OVERVIEW OF THIS BOOK

This volume provides an introductory overview of policy and management strategies for climate adaptation. As a starting point, we accept the scientific consensus around the causes of anthropogenic climate change, the present-day manifestations of this change, and the projected future pathways that uniformly suggest increasing impacts associated with this change. Our primary topic here is the ways human societies can and should respond to those impacts. We explore policies—national and subnational—and also consider the roles of nonstate actors like civil society and the private sector; the result is an approach that considers the *governance* of climate change adaptation across sectors and geographies.

Columbia University's Earth Institute Sustainability Primer series offers a forum for contextualizing aspects of sustainability for its intended audience. This book is targeted at academic and government audiences as well as the general public, and readers without a technical background or advanced degree should find it accessible. It is intentionally short, and the language is light on unexplained technical jargon. Likewise, disputes over terminology and theory within the academic literature are not deeply probed here. Those who seek more in-depth analysis may begin by exploring the comprehensive reference list at the end of the book; it's also worth noting that this is a dynamic and vibrant field of study with rapidly updated content, and this volume may not always represent the most current scholarship.

The book is organized as follows. To orient readers, this introduction presents key terminology and describes connections among foundational concepts. Chapter 1 then provides an overview of contributions from both the natural and social sciences to the growing field of climate adaptation while also taking a deep dive into relevant institutional and governance structures. Since many of those institutions are focused on reducing risks associated with climate-driven extreme events, chapter 2 explores disaster risk reduction (DRR), a long-standing and vibrant field that has recently begun to merge with climate adaptation efforts. Given the prevalence of infrastructure as a risk reduction tool, chapter 3 considers the ways infrastructure-based strategies have been used historically and are now being modified to improve adaptation outcomes. Chapter 4 goes deeper into the urban setting, examining how dense networks of human settlements can address rising temperatures through a discussion of zoning, building codes, and other planning efforts. Refocusing outward from cities, chapter 5 covers dimensions of adaptation more associated with rural areas, including land use, agriculture, and food security, and highlights patterns of vulnerability in the farming sector. From there, chapter 6 shifts the focus to the emerging role for insurance as a tool to buffer farmers, communities, and governments from the impacts of a changing climate. When this array of dynamic and interconnected adaptation strategies—DRR, infrastructure development, urban planning, modified land use, and insurance—does not succeed in protecting people, they may be forced to migrate in search of better living conditions. Chapter 7 takes a closer look at climate-driven displacement and migration patterns, giving attention to ways in which an adaptive response might maximize benefits from those movements. Throughout the book, themes of equity and justice frame key issues, and in chapter 8, those strands come together to provide a more coherent analysis of inequality. Chapter 9 summarizes important lessons learned and next steps for research, application, and study.

ADAPTING TO WHAT?

Scientists have long pointed to the correlation among emissions of GHGs, the rise in concentrations of carbon in the atmosphere, and increases in global temperatures. Since its creation in 1988, the Intergovernmental Panel on Climate Change (IPCC) has been the scientific body charged with providing peer-reviewed, updated, and consistent assessments for policy makers. Every few years it releases updated scientific reports, including the work of Working Group II on impacts, vulnerability, and adaptation. As of this writing, the most recent published findings come from the *Fifth Assessment Report*, published in 2014. In that report, scientists conclude unequivocally that anthropogenic—human-caused—climate change has already caused measurable impacts on both natural and human systems. Shifting precipitation patterns mean some areas are experiencing more rain than historical averages while others contend with drought. Melting glaciers also contribute to hydrological changes, including unstable permafrost and increased runoff. Terrestrial and aquatic species are facing changes in habitat conditions, leading them to alter their migration patterns and contributing to declining population numbers. Agricultural producers are seeing reduced yields in some areas and increased yields in others. Extreme weather events have increased in frequency, intensity, and duration. Human mortality due to heat waves portends greater threats to human health in the future. Scientists note that these environmental and human impacts also interact with each other, creating a web of positive and negative feedback loops (figure o.1). Cumulatively, impacts are projected to worsen as the global temperature continues to rise, and the IPCC projects warming in excess of 2°C by the end of the twenty-first century.[1]

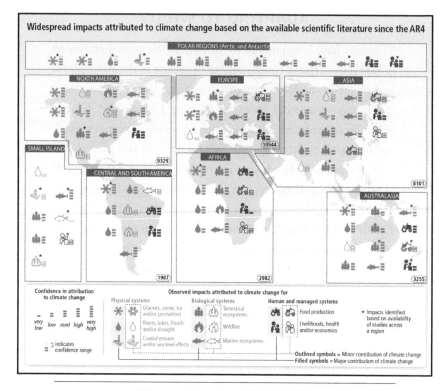

FIGURE 0.1 Widespread impacts attributed to climate change.

Source: Figure SPM.4 from "Summary for Policymakers," in *Climate Change 2014: Synthesis Report: Contribution of Working Groups I, II and III to the Fifth Assessment Report of the Intergovernmental Panel on Climate Change*, ed. R. K. Pachauri and L. A. Meyer (Geneva: IPCC).

Climate modelers have developed sophisticated tools for projecting future impacts within a defined range of probability. Indeed, climate science is robust, and technical expertise is high; still, translating global data into information useful at a national or local level—where adaptation policy responses are crafted—can be challenging. For example, we know that global temperatures are expected to rise, but those increases are not evenly allocated across the globe. Some areas will see (and are already seeing) dramatic temperature changes while others are less likely to experience them.

Further complicating the adaptation challenge is the reality that a community seeking to adapt to climate change must consider not only immediate, proximate impacts but also distant impacts with localized ripple effects (figure 0.2). Sea-level rise may affect coastal communities directly but also impact inland and downstream communities indirectly. One country may find it is experiencing both floods and droughts in different geographical areas, leading to low yields for some crops and yield increases for others. Those variations are then reflected in export prices and relative food availability in the grocery stores of trading partners around the world.

Indeed, identifying risks at regional, national, local, and individual levels may require different modes of data collection. Matching the scale of the projected impacts with the scale of the policy and management response is essential; in other words, climate change impacts projected at a national level will be of little use to a city planner who is seeking to implement adaptation measures for local residents. But these layers are all connected through space and time, as local impacts are embedded in an international eco-social system.

Unpacking these scientific intricacies is further complicated by the ways many such hazards interact with one another to create compound risks. Scientists who study future climate risk derived from these compound events have seen how interactions among multiple drivers of climate change can amplify impacts on the ground. The co-occurrence of multiple hazards is particularly challenging for scientists to dissect, as "unusual combinations of processes associated with the events makes them difficult to foresee."[2] For example, a simultaneous heat wave and drought vastly

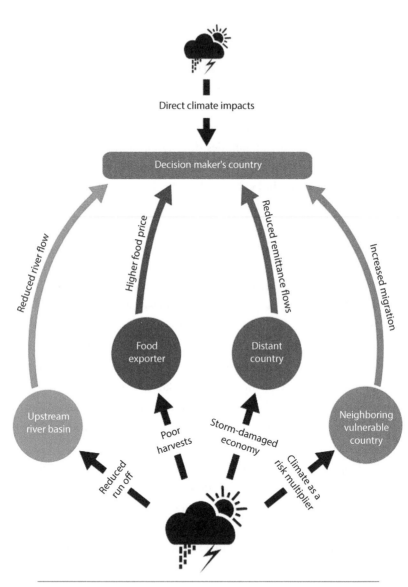

Direct climate impacts

Decision maker's country

Reduced river flow

Higher food price

Reduced remittance flows

Increased migration

Upstream river basin

Food exporter

Distant country

Neighboring vulnerable country

Reduced run off

Poor harvests

Storm-damaged economy

Climate as a risk multiplier

FIGURE 0.2 Impacts from climate change are both direct and indirect.

Source: Image adapted from M. Benzie, "National Adaptation Plans and the Indirect Impacts of Climate Change" (SEI Policy Brief, 2014).

increases the risk of dangerous wildfire. All three of these hazards can be measured separately, but taken together, they amount to compound risk that can elude quantification.

Social scientists remind us that human societies are not just passive recipients of climate change impacts. Humans contribute to rising temperatures through their use of fossil fuels and their participation in a carbon-intensive economic system; they also participate in the framing of risk and in society's response to that risk. It is there that the study of governance of climate change adaptation must begin.

TERMINOLOGY

Like any technical field, climate adaptation is full of jargon. In addition to scientific language, which can be incomprehensible to a layperson, several key terms are used conversationally and have the potential to be misunderstood. These terms are used in high-level international conversations, in local decision-making contexts, and throughout informal community interactions. Establishing a consistent meaning for each term is thus challenging and, in some cases, place based. In other words, "the linguistic differences are not just in translation and interpretation, but are also cultural."[3] Those cultural filters result in "circuits of meaning" that have implications for policy implementation and decision-making at all levels.[4] Thus, while the IPCC offers a standardized definition for each term—provided here—more nuanced connotations and meaning are also essential for deeper understanding.

Hazard: the potential occurrence of a natural or human-induced physical event or trend or physical impact that may cause loss of life, injury, or other health impacts, as well as damage and loss to property, infrastructure, livelihoods, service provision, ecosystems, and environmental resources.

When referring to a distinct event or process unfolding over time, we usually understand a hazard to be an external threat. Note that this definition is agnostic as to its driver; that is, whether or not the dangerous event is anthropogenic is, from a definitional standpoint, irrelevant. In this way, the term is relatively free of normative implications. Hazards are neutral. But we must also consider the ways in which our very notion of a hazard might be socially constructed. For example, if a wildfire takes place in a remote wilderness area, we might not consider it a hazard, as human assets are not subject to harm. If the fire shifts direction when wind patterns change and suddenly threatens human lives or property, that same force of nature is interpreted differently and will likely be labeled a hazard. There are also temporal dimensions to the concept. We can easily identify a rapid-onset hurricane approaching a major coastal city as a hazard and respond with urgent action. Slow-onset hazards, like droughts, tend to be harder to categorize. The parable of the frog being slowly boiled alive reminds us that it can be difficult to identify threats under certain contextual conditions. In this way, rather than seeing a hazard as a purely objective event, we might reconceptualize the term to refer to "phenomena with properties that society can make hazardous."[5]

> *Exposure*: the presence of people, livelihoods, species or ecosystems, environmental functions, services, and resources, infrastructure, or economic, social, or cultural assets in places and settings that could be adversely affected.

This term is used conversationally to signify an unprotected asset, left to whatever the elements may bestow. If hazards are seen as external threats that humans cannot directly remove, exposure is something humans do have an ability to manipulate. In the wildfire example, a home that is situated in a heavily wooded area would be seen as highly exposed. As the hazardous wildfire approaches, exposure will make it more likely that the house will experience harm. Modifying the property in certain ways—like clearing brush near the home before it is threatened by a fast-moving fire —would do little to affect the hazard but could have meaningful implications for the home's exposure.

Vulnerability: the propensity or predisposition to be adversely affected; it encompasses a variety of concepts and elements including sensitivity or susceptibility to harm and lack of capacity to cope and adapt.

Vulnerability is perhaps the most nuanced term listed here. In general, it refers to longer-term and systemic patterns of susceptibility to damage, amplified through exposure to hazards. Returning to the wildfire example, if the residents of the exposed home in the path of that hazard are also unemployed, without substantial savings, uninformed about wildfire risk , and without home insurance, we might consider them to be highly vulnerable to the hazard. They face long-term and severe risks to their livelihood if that fire damages their home. Alternatively, if those homeowners are informed about wildfire risk, have insured their home, and have resources to relocate in the event of being displaced, we would likely consider their vulnerability to be relatively low. The same hazard in a home with the same level of exposure will have dramatically different effects on the individuals living there based on their vulnerability status. In this way, the very notion of a disaster becomes inextricable from the concept of vulnerability.

> *Risk*: the potential for consequences when something of value is at stake and where the outcome is uncertain, recognizing the diversity of values.

The notion of risk draws on assessments of hazard, exposure, and vulnerability to offer an integrated understanding that can drive behavior. Scientists have long sought to quantify the concept of risk, and, indeed, assigning meaningful numerical values to components of risk is the beating heart of the insurance industry (see chapter 6). Even mainstream scientific publications refer to risk using formulas that define it as a mathematical function of hazard, exposure, and vulnerability (figure 0.3). But critics have noted that any such formula amounts to a "pseudo equation"; they argue that layers of complexity confound risk methodology, particularly when applied to multidimensional

RISK = HAZARD x EXPOSURE

FIGURE 0.3 Modifying exposure can significantly reduce risk.

Source: European Plasticisers.

socioeconomic systems.[6] In part, the challenge stems from the fact that historical averages, so often used to predict future trends and quantify departures from those trends, no longer provide solid benchmarks in an era of climate change. Instead, climate change scientists are attuned to the reality that "for every incremental change in greenhouse gases and temperature, there are diverse responses in climatological, ecological, hydrological, and other biophysical systems, ranging from short-term effects on primary productivity, through to longer-term changes such as sea-level rise, weather or soil formation; coupling of systems means that responses also affect other systems, including feedbacks to climate."[7] Understanding how multiple risks cascade through interlocking systems is a scientific challenge that renders unlikely the identification of a fixed, quantifiable measurement of risk.

Adaptation: the process of adjustment to actual or expected climate and its effects. In human systems, adaptation seeks to moderate or avoid harm or exploit beneficial opportunities.

In some venues, adaptation is defined only in opposition to mitigation. That frame sees reducing GHGs as the primary goal and regards adaptation as something that happens only when mitigation has failed. More recently, scientists have begun to see that "adaptation and mitigation are not a binary opposition; they influence each other in positive and negative ways."[8] When that relationship is negative—when an action that benefits mitigation goals contributes to poorer adaptation outcomes, for example—practitioners must navigate complex trade-offs. When that equation is positive—when a single action benefits both mitigation and adaptation goals, for example—we call the results *co-benefits*.

Approaches to adaptation range from incremental to transformative. The dominant policy approach is incremental, mostly government led, top-down, and amenable to technical solutions. Decision makers turn to existing institutional structures to do business a little bit differently. Pervasive in the system led by the United Nations (UN), these incremental approaches tend to rely on data from climate models to assess risks, and they seek win-win solutions to minimize those risks. For example, in an effort to reduce costs associated with climate-driven extreme weather and storm surge, many national and subnational governments have added requirements for developers wishing to build on a coastline. This approach layers regulatory components onto long-standing process requirements but fails to rethink the paradigm. Developers may be required to navigate new layers of environmental review that include planning for predicted climate conditions or obtaining property insurance in a flood zone. The institutions will still be familiar to the developers, and the process is largely unchanged. In the end, perhaps they will have to make some modifications to the site plans, but they will ultimately be permitted to build on the coast.

Incremental approaches tend to be both economically and politically feasible. They offer an important early pathway for policy adjustments that address risk. But they don't address the sociopolitical root causes of vulnerability. As we move along the continuum toward more transformational change, the need for societal reform to reduce vulnerability and increase capacity becomes central to adaptation planning. Rather than prioritizing the addition of new layers of review for coastal development, for example, decision makers seeking transformational adaptation would pursue more comprehensive solutions to coastline development. They would also examine the socioeconomic condition of exposed communities nearby and propose broad capacity-building measures that would better equip those residents to be agents of change. Ideally, their process would engage those residents in participatory governance leading toward a new pathway featuring climate-smart development; under these conditions, the coastal development proposal would likely be denied.

Adaptive Capacity: the ability of a (human) system to adjust to climate change (including climate variability and extremes), to moderate potential damages, to take advantage of opportunities, or to cope with the consequences.[9]

A governance approach centered on capacity building is considered by many scholars to represent the most important target for adaptation policy, as a community with strong adaptive capacity will be well positioned to proactively engage in risk minimization, no matter what form those risks take. This approach focuses less on infrastructure and short-term problem-solving and more on long-term, intergenerational transformative change. It is rooted in the belief that current socioeconomic systems are fundamentally flawed and inequitable and that relying on incremental reform can unintentionally serve to reinforce those systems. The key argument is that a deep understanding of adaptation must include the notion of structurally driven vulnerability. In this way, adaptation is framed more as a process than an outcome. And while quantitative risk modeling is still essential for building adaptive capacity, those data render invisible the "web of enabling and constraining factors, economic incentive structures, different knowledge systems, and diverse patterns of environmental governance."[10] Additional sources of information must be added to quantitative risk data in a robust adaptation framework.

Resilience: the capacity of social, economic, and environmental systems to cope with a hazardous event or trend or disturbance, responding or reorganizing in ways that maintain their essential function, identity, and structure, while also maintaining the capacity for adaptation, learning, and transformation.

The terms *resilience* and *adaptation* are often used interchangeably by journalists and scholars, but in fact, they refer to different dimensions of the same concept. The former comes from the Latin *resilire*, meaning "to jump back," and when the concept emerged in psychology in the 1950s, it was used primarily to describe the human ability to respond to stressors. In the 1970s, ecologists began using the term to describe ecosystem function. By the 1990s, it was more extensively used in the DRR field, and systems thinking had embedded it in the notion of socioecological resilience. Now resilience is understood to be normative and generally signifies the relative presence of adaptive capacity, although that capacity may be understood as a system trait, a process, or an outcome.[11]

Scholars have observed that this term may suffer from timescale confusion because sometimes short-term resilience can contribute to longer-term catastrophe. For example, if a community's resilience allows it to bounce back from short-term shocks, it may enable that same community to avoid taking more transformational long-term action. One scientist has described herders in Ethiopia when a famine hit.[12] Many sold their most valuable assets so they could afford food. The money they earned allowed them to persevere through the famine's early months. That's resilient! But as time went on and the famine didn't resolve, they found themselves broke and stranded. The resilience they had deployed to deal with the short-term shock actually served to undermine their long-term adaptive capacity. In these ways, the notion of resilience has long been interwoven with patterns of vulnerability, and the two terms "were used to describe the choices available to people and ecosystems in response to . . .

natural events well before discussion of climate change began."[13] In this book, we use the term narrowly to indicate efforts that strengthen existing settlement and livelihood patterns.

> *Maladaptation*: action or inaction that may lead to increased risk of adverse climate-related outcomes, increased vulnerability to climate change, or diminished welfare, now or in the future.

In earlier reports, the IPCC defined *maladaptation* as adaptation that does not reduce vulnerability but instead increases it. Since then, scholars and practitioners have expanded the notion to involve an array of outcomes that may undermine collective efforts to manage risks associated with climate change. Embedded maladaptation can begin in the planning phase, is itself a process (rather than an outcome), and has both spatial and temporal dimensions. Connections to vulnerability are particularly noteworthy. Both terms—*vulnerability* and *maladaptation*—refer to dynamic processes that interact with one another over time.[14] Five definitional components of maladaptation have emerged in the literature as the most salient: action that leads to increased GHG emissions, a disproportionate burden on the most vulnerable, high opportunity costs, reduced incentives to adapt, and/or increased risk of path dependency.[15]

Building seawalls is a classic example of maladaptation. Many coastal cities have addressed the risks from sea-level rise and storm surge through the construction of seawalls. But this strategy hits all five criteria of maladaptation. First, a seawall is usually constructed of emissions-intensive concrete and built with heavy equipment powered by fossil fuels, contributing to the very problem it seeks to address. Construction-related emissions will lead to faster and more intense sea-level rise! Second, the water that is blocked by the new seawall will still come ashore somewhere, now most likely downstream where vulnerable communities without the financial ability to build their own seawall lie. In this way, the wall contributes to an intensification of vulnerability in those lower-capacity areas. Third, money devoted to the expensive seawall could have been more productively deployed to other, more-transformational adaptation efforts; opportunity costs are high. Fourth, residents of the city that is now protected by the new seawall are likely to consider themselves safe from coastal threats. They will be less inclined to support more transformational adaptation measures. Finally, the construction of one seawall sets into motion a development pathway. Technical experts will populate institutions created to build and maintain that wall, budgets will be reconfigured to provide funding, and as the seas continue to rise, that city is quite likely to add layers to its seawall, committing to an unsustainable pathway for development.

This overview of key terms should empower readers to take the next step into foundational scientific and governance elements of climate change adaptation. In chapter 1, we apply the terms introduced here to explore connections between science and policy and to consider roles for the most salient institutions designed to address adaptation.

1

FOUNDATIONS

Science, Policy, and Institutions

CONNECTING SCIENCE AND POLICY

Scientists are making more precise global climate projections and are increasingly able to attribute extreme weather to anthropogenic climate change, and yet the resulting data may offer little in the way of guidance for action at the local level. How can a rural farmer in Bangladesh or a small-town mayor in Iowa or an urban planner in Tokyo translate the likelihood of global sea-level rise into something meaningful for their community? For the moment, let's put aside the obvious communication gaps in play at the individual level—that Bangladeshi farmer, for example, isn't reading the scientific reports—and consider how decision makers at all levels obtain updated and actionable scientific information. How does scientific knowledge translate into adaptation policy?

While evidence-based decision-making is the gold standard for linking science and policy, its application presents a range of challenges. First among them is the inequitable access to scientific information. Some countries with high scientific capacity— including European Union (EU) member states, the United States, and some states in Asia—have developed sophisticated decision support systems for gathering information and providing guidance to practitioners. Some of these systems offer guidelines for local officials, some compile information that connects projected impacts with adaptation options, and some offer detailed instructions for conducting risk assessments. The most well-developed of these systems operate through online portals where practitioners can report their monitoring results and find solutions that match their circumstances. The EU's Climate ADAPT tool is one of the most robust examples to date (figure 1.1). The first versions of these adaptation support tools were developed in the early 2000s, but most are newer and haven't yet been evaluated for long-term effectiveness. In 2018, researchers found that among eighty-eight such tools they reviewed, only five had been systematically assessed.[1]

Less wealthy countries often lack these layered portals for scientific knowledge building. They may have additional capacity weaknesses, such as a shortage of trained scientists and inconsistent access to the internet, that present barriers to obtaining and making sense of raw scientific data. *Climate services*, defined as the provision of scientific information about climate change to assist decision-making, are scarce. This issue has been addressed in some locations through UN agencies, academic institutions, and nonprofits that conduct targeted outreach, but it remains a serious problem in many of the hardest-hit countries. For more

FIGURE 1.1 The EU's Climate ADAPT online tool offers country profiles, EU sector policies, case studies, and an interactive adaptation support tool.

Source: European Environment Agency.

discussion of how climate services can foster an enhanced adaptive response through index insurance, see chapter 6.

Scientific uncertainty, even in the most well-provisioned countries, continues to undermine adaptive action. It is perhaps a feature of the human condition that we find it so difficult to act when the location and timing of future hazards are known only through probabilities. We are more naturally attuned to short-term planning horizons and tend to employ cognitive discounting heuristics—unconscious preferences that value the same reward more highly when it is delivered now rather than later—when projecting current scientific information into the future and across scales.[2] Slow-onset impacts are particularly difficult for humans to detect or understand. Uncertainty contributes to political hesitation and has fueled the fire of climate denialism. The most appropriate response to uncertainty may emerge from the same basic principle that underscores much of environmental policy: the *precautionary principle*, which suggests that even in the absence of surety, when the consequences are potentially dire, action is warranted. The issue of climate adaptation also has normative dimensions. We may believe that taking action is something a responsible government *should* do, even in the absence of certainty, as minimizing vulnerability serves to benefit society writ large.

Even when the science is clear, the information is available, and the decision makers are equipped to both understand it and take action, policy may not follow. After all, "policy-making is not primarily about solving problems based on scientific evidence, but rather a messy political power struggle between different types of evidence, values, ideologies and economic interests held by a broad variety of actors."[3] For these reasons, we shouldn't mistakenly view the science-policy nexus as a linear pathway where science leads inexorably to policy. Rather, the needs and priorities of local decision makers, the political context within which they operate, and the resources available to devote to the issue result in powerful intervening forces that must be negotiated on the journey from science to policy. Political scientists are quick to remind us that the pathway is not unidirectional; policy also drives scientific inquiry through funding allocations, agenda setting, and a range of other political and social factors.

As an interim step in this long-term process of creating adaptive policy, *climate risk management* (CRM) connects real-time science with short-term management changes. The notion is that some actions might be politically palatable even when they don't address fundamental structural changes that may be needed to advance transformational change in the distant future. Identifying immediate hazards and seeking to reduce exposure can succeed in mitigating risk at least in the short term; more importantly, these incremental steps can begin to institutionalize pathways along which scientific information can travel from technical isolation, through intermediary institutions, and ultimately to the policy-making sphere.[4]

INSTITUTIONS AND GOVERNANCE

Institutions—variably understood as brick-and-mortar formalized organizations or as more nebulous established practices and customs—play an essential role in climate adaptation governance. Inflexible institutions can create path dependency, erecting invisible barriers to even incremental new policy approaches. Flexible, adaptive institutions, by contrast, can boost efficient implementation of new policy streams and react quickly to new information. They are central to two related concepts: policy and governance. The first term refers primarily to actions taken by governments.

At all levels, *public policy* is created by elected or appointed public officials. It is funded through public sources of money like taxes, implemented by civil servants in a bureaucratic structure, and enforced by legally recognized authorities. But plainly not all actions that might matter for climate adaptation are undertaken by governments alone. Civil society and the private sector play important roles, and those efforts can serve to incentivize behavior change in coordinated ways. These broader actors contribute to *governance*. Here we will explore how this variety of actors, both public and private, functions in the emerging climate adaptation regime.

Global Institutions

In the early years of UN-led global governance for climate change, misguided hope led many to conclude that mitigation efforts would be effective, and adaptation was viewed as both more abstract and more temporally distant. Even the UNFCCC, created in 1992 to facilitate international dialogue on the matter, identified a relatively narrow mission: "stabilization of greenhouse gas concentrations in the atmosphere at a level that would prevent dangerous interference with the climate system."[5] Perhaps it's no surprise that climate change action in the early years was almost entirely focused on mitigation.

The first global treaty to address climate change, the Kyoto Protocol (1997), was enacted in the midst of intense global controversy about its merits. Its placement of countries into one of two categories designed to measure their relative wealth—Annex I for wealthier, industrialized countries and non–Annex I for everyone else—caused the most criticism. Some Annex I countries argued they were being held to unfair binding emissions targets that didn't apply to non–Annex I states, potentially hobbling their own competitive economic growth. The United States found these concerns so salient that it declined to ratify the treaty. Notably, the Kyoto treaty process launched the Adaptation Fund (AF), but only developing countries, which were seen as having little role to play in global mitigation efforts, were asked to think about adaptation. At the time, the adaptation needs of industrialized states received very little global attention.

In the late 1990s and early 2000s, non–Annex I states were encouraged to develop *National Adaptation Programs of Action* (NAPAs). With financial support available through the UN's Least Developed Countries Fund and technical support provided by the UN Development Program (UNDP), many countries took up the challenge, working to identify the most urgent and immediate climate threats facing their societies and mapping out prioritized activities to be taken in response. But without a global mandate for action, most countries did not prioritize completion. As of January 2021, only thirty-one countries had completed these plans. By making the finished NAPAs widely available, the UNDP hoped to foster peer learning and build global capacity for action. But with the focus only on short-term hazards, NAPAs better align with a CRM approach and do not function well for longer-term strategic planning. They are now seen as an important, if incomplete, first step toward comprehensive adaptation planning.

By 2007, with Kyoto starting to lose momentum and climate impacts becoming readily apparent, the IPCC amplified the need for all countries, even those designated Annex I, to focus on adaptation. The Bali Action Plan, launched that year at the Conference of the Parties to the UNFCCC (COP13), proposed a detailed approach to the challenge, and in 2010, the Cancun Adaptation Framework offered even more robust guidance.

The UNFCCC's Adaptation Committee was formed a few years later at COP16, and by the following year, countries were invited to begin developing National Adaptation Plans (NAPs). These plans built upon earlier NAPA efforts but now have an explicit focus on long-term adaptation needs with three key objectives: reduce vulnerability, build adaptive capacity, and facilitate integration of adaptation into relevant policies. The NAP Global Network was created at COP20 in Lima, Peru, to provide support for countries building their plans, and since then, it has grown into a flourishing clearinghouse of scientific information, planning templates, case studies, and technical aid. To date, 143 countries have participated in the network, 40 of which have received direct technical support.

By the time the Paris Agreement (PA), was signed in 2015, it was clear to many that mitigation efforts were insufficient. Climate change was happening, and adaptation was taking root, both formally and ad hoc. Structured with a new, bottom-up model for global treaty development, the PA marked a turning point in the governance of climate change. Each country submitted its own *intended nationally determined contribution* (INDC), describing in detail how they would help to maintain global warming below 2°C. Every INDC has a section on adaptation, including an overview of projected hazards and a road map for developing strategic policy responses. When the treaty went into force less than a year later, the word *intended* was dropped, and the *nationally determined contributions* (NDCs) became self-defined binding pledges.

Article 7 of the PA set out twin global goals for adaptation: enhancing adaptive capacity and reducing vulnerability. The PA also created new reporting instruments for countries to use to share information, and a range of technical and financial support instruments began to creak to life. Since the passage of the PA, global institutions devoted to the adaptation challenge, as well as those focused on adjacent and overlapping tasks like sustainable development and disaster risk reduction (DRR), have proliferated and expanded their mission to incorporate adaptation priorities.

Three separate funds that can be tapped for adaptation support—the Global Environment Facility (GEF), the AF, and the Green Climate Fund (GCF)—have been established over time through the UN system, each with different sources of funding and different prioritized uses. When early results amplified equity concerns, separate working groups on indigenous rights, gender dimensions, and the particular needs of less developed countries (LDCs) were established. Those important issues are discussed at greater length in chapter 8.

Critics will note that a proliferation of global institutions doesn't necessarily correspond with substantive progress; still, the creation of high-level work bodies devoted to adaptation suggests widespread awareness of some of the nuanced challenges of implementation. But this international framework cannot do the heavy lifting. Only sovereign states, with support from global institutions, the private sector, and civil society, can develop, implement, and enforce new policies.

National Governments

National governments play a lead role in agenda setting and prioritization for climate change adaptation. Research has shown that the most daunting barriers to effective adaptation include lack of awareness, scientific uncertainty, lack of resources, and weak political commitment; national governments are best positioned to address all of these.[6]

As a starting point, they are charged with implementing domestic laws that will achieve targets set through the international treaty process. Indeed, most national governments now have climate mitigation policies, adopted in part as an effort to achieve targets set by the 1997 Kyoto Protocol and later updated and reinforced by the 2015 PA. Those policies include both statutory and regulatory components and may, for example, require the power sector to shift toward more renewable sources of energy or the transportation sector to include more efficient vehicles. Regional suprastate institutions like the EU have worked to standardize some of these national policies across boundaries so as to reduce friction for the private sector. But so far, attention to adaptation has been slower to develop and is less coordinated. Adaptation is frequently seen as a local responsibility that eludes a national statutory fix. And while strategies such as carbon taxes and cap-and-trade schemes have proven useful for mitigation, the challenge of adaptation doesn't lend itself readily to a market solution. So what can national governments do to advance adaptation measures?

Three broad categories of policy instruments are available for national governments. First, legal approaches, including policy making through statutes and regulations, form the backbone of a national government's toolbox. Policy wonks tend to divide environmental policy into two categories based on the lever used to drive behavior change: command-and-control and market-based strategies. Historically, most environmental policy making followed a command-and-control template. Governments established pollution standards, such as limitations on how many particulates of a dangerous chemical industries were allowed to release, and violations of those standards were met with fines or other enforcement methods. In recent decades, governments have moved to market-based approaches, such as cap-and-trade programs and carbon taxes, a shift that has gained favor for what is seen as the increased flexibility it gives to industry. Rather than mandate a shift to cleaner production methods, for example, a market-based policy establishes financial rewards for a company that does so.

But climate change adaptation cannot be governed exclusively through either of those approaches. The goal for adaptation can be elusive, often referring more to a process over time than an identifiable outcome, and, therefore, the practice isn't well suited for traditional policy approaches. Still, reforming existing policies can be a potentially powerful adaptation tool for national governments. Breaking down long-term process ambitions into shorter-term operational tasks can help. Many countries require environmental review before approving large infrastructure projects, for example, and those laws can be updated to mandate consideration of climate change impacts. National agriculture policy can powerfully drive farmer behavior, and an adaptive framework could redirect subsidies to support climate-smart crops. The provision of social support services, including free and accessible public education and health services, can similarly be amplified to better build adaptive capacity.

Second, national governments have uniquely powerful platforms for sharing information. Although research suggests awareness of climate change is growing everywhere, most of that knowledge is related to mitigation, and there is persistently less understanding of adaptation.[7] National governments have an important role to play in this area, including serving as a clearinghouse for scientific information, building public awareness, setting a national agenda, and framing key issues. Deploying environmental ministries to translate technical scientific information for the broader public is essential, and providing a trusted venue for the provision of updated science can help to reduce the spread of misinformation. Signaling a political commitment to climate adaptation has the potential to trickle down to local units. Some governments have succeeded in this.

Third, national governments have authority over fiscal policy and resource allocation. Through the prioritization of funds, these governments powerfully shape what's possible by other layers of government. Budgetary decisions have profound effects on climate adaptation. More discussion on the financial dimensions of adaptation can be found in chapter 8.

In many instances, countries are already pursuing a range of strategies to reduce poverty, spread risk, and plan for disasters. These policies, while likely not initially framed as adaptation, contribute meaningfully to a robust adaptation regime, as they can serve to strategically reduce vulnerability. The challenge for some countries, then, is to organize disparate adaptation measures, identify redundancies and gaps, and develop a coherent prioritization strategy.

Some national governments have approached the adaptation challenge as a call to better empower their local counterparts. To do this, they have focused on creating an enabling policy framework at the national level and then have provided both funds and authority for decentralized governance. In those settings, the establishment of tracking and reporting mechanisms can help ensure accountability. Policy coherence also means national frameworks and local action should align with global guidelines. This level of integration is still weak across much of the world and is ultimately a pivotal task that only national governments can perform.

Local Governments

Even in countries with strong political will for enacting adaptation measures, centralized policy making will not be sufficient without robust partnerships at the local level. Ultimately, successful climate adaptation relies on centralized governments to develop policies that empower and equip local leaders; unlike mitigation, which is unavoidably a global collective-action problem requiring broad participation, adaptation is often seen as "largely local and [requiring] little international coordination."[8] Impacts will be experienced locally, and actions designed to minimize those impacts must similarly be enacted at the local level. Even with robust national enabling legislation, globalized trade policies, and international institutions that seek to standardize the practice of adaptation, localities are at the forefront of the action.

The focus on local policy-making aligns with two key normative principles of economics: the subsidiarity principle and the correspondence principle. Asserting that decisions should be made by the decision unit at the lowest level of aggregation possible, the *subsidiarity principle* would devolve adaptation action to the most decentralized unit that can handle it. The *correspondence principle* seeks to ensure that beneficiaries of a good are the same economic actors who both bear the costs of managing that good and have the authority to make decisions regarding equitable allocation of that good. In other words, even when a national government might be capable of adopting uniform adaptation strategies, it shouldn't. Those decisions should be reserved for local governments, which are better positioned to respond to localized threats and develop locally feasible policy responses that empower the same individuals who experience direct harm from climate change.

But devolution of policy authority comes with risks. Significant barriers to effective local government action include weak administrative capacity, local political pressures that may prioritize other policy arenas, limited financial resources, lack of scientific understanding, and difficulties with monitoring, reporting, and enforcement. In sub-Saharan African countries, for example, researchers found that local governments are largely at the mercy of central governments for authorization and implementation of resource policies; the result is that front-line local authorities "lack vital discretionary powers over most adaptation domains."[9] Even when national governments defer to local entities, allowing them to take the lead in governing, the outcomes are only as strong as those local units are. Wide disparities in capacity can have disastrous results, including intensified vulnerability and increased exposure.

Civil Society

Despite the prevalence of UN-administered public institutions in global governance and the centrality of governments in all dimensions of adaptation policy making, organized individuals have a powerful role to play. Civil society has blossomed in the adaptation space. Transnational and domestic civil society organizations (CSOs) bring knowledge, accountability, and critical public support to many climate change activities. For example, the Carbon Disclosure Project (CDP), well known for its groundbreaking work in tracking GHG emissions by sector and country, makes its massive collection of data available for free. With information on more than 800 cities in 120 countries, the CDP can help to fill capacity gaps that hobble many local governments. Similar CSOs have sprung up in countries around the world, serving as a bridge between research institutions and the public. Advocating for vulnerable populations, CSOs have also helped to amplify the adaptation needs of marginalized communities that might otherwise lack direct access to the political process. Climate justice and human rights CSOs have increasingly incorporated an explicit focus on adaptation, contributing to building resilience, promoting awareness, and mobilizing people to push their governments for better governance solutions.

The Private Sector

The private sector can also serve in a governance role. Corporate sustainability has grown significantly in the last decade, with nearly every large company now equipped with a sustainability office and a public relations machine to publicize emissions reductions, waste stream improvements, and climate change practices. The public has responded by voting with its collective wallet, choosing to consume products that come from greener organizations.[10] But most of those corporate efforts are linked to broad mitigation goals, and less attention has been paid to adaptation dimensions. To date, companies have not forged a clear connection between their products and the adaptation challenge faced by local communities.

A similar pattern emerges for foreign direct investment (FDI). A powerful tool in sustainable development for impoverished countries, FDI is a foundational building block for international cooperation, as it can provide stable and durable links that transcend politics. It has been powerfully deployed to shift incentives for climate change mitigation in some instances; indeed, over two-thirds of the costs associated with climate mitigation are borne by the private sector.[11] This is not the case with adaptation, where only 8 percent of total financing for adaptation has come from private sources. According to the Global Adaptation and Resilience Investment Working Group, "uncertainty and a perceived lack of investible opportunity have limited . . . private investment in adaptation and resilience to the physical risks of climate change."[12] But the future may open doors. Fully 93 percent of surveyed organizations with investment portfolios reported an interest in such opportunities within the next three years. Those would-be investors suggest better outcomes would result if there were more consistent regulatory standards, better disclosure of investment risk, and more opportunities connected to blended finance vehicles.

While corporate and investment finance flows have not meaningfully altered the adaptation landscape, the private sector does have a critical role to play through the insurance industry. As governments, businesses, and individual farmers increasingly seek out opportunities for risk pooling as a strategy to buffer themselves from the effects of extreme weather, the insurance industry has responded with new tools and more accessible products. The complexities of insurance provision at all scales are explored in detail in chapter 6.

DEFINING SUCCESS

Unlike climate mitigation, which is tracked through a range of clear numerical targets including an inventory of GHGs by sector and country, adaptation efforts lack a set of shared indicators for success. Indeed, "in the discourse on climate adaptation there is an acknowledged lack of consensus regarding what exactly constitutes successful and/or sustainable adaptation actions."[13] Part of the challenge lies in the difficulty of measuring a counterfactual:

e.g., if this seawall hadn't been built, the city might have flooded. Another part of the challenge is the dependent variable problem, which refers to conceptual confusion about what is being compared between cases.[14] Are we mostly concerned with risk reduction? Or vulnerability? Or capacity building? Difficulty with measuring success leads to the deep uncertainty that is a hallmark of a *wicked problem*. Such a problem, by definition, defies a clear solution; solving it means balancing trade-offs and accepting some harm in pursuit of some good. But how are those trade-offs measured and assessed?

Quantitative methodology, common in the natural sciences and economics, can capture key measurements that are useful for cost-benefit analysis (CBA), historically a pivotal tool for making choices among competing policies. For example, changes in the market value of key assets might reveal shifts in risk profiles. Disability-adjusted life years (DALYs) can help to tease out potential impacts on public health as different climate policies are enacted. Wealth saved might reveal ways in which policies modify vulnerability patterns and poverty.

Some of these tools have already provided useful insights for climate adaptation. For example, the UN Institute for Environment and Human Security now offers a decision support tool called the Economics of Climate Adaptation (ECA). Based on the climate-modeling tool CLIMADA, the ECA develops a risk map based on hazard simulations, identifies spatial distribution and values, estimates damage functions, integrates vulnerability, and then recommends different adaptation functions, ranked according to efficiency. This tool has been made public and can be used to help countries build their NAPs. In another example, the Global Commission on Adaptation, co-led by the World Resources Institute and the UN Global Center on Adaptation, in 2019 offered cost-benefit ratios for investments in different adaptation dimensions. One finding was that strengthening early warning systems (EWSs) has the highest ranking for benefit, with a cost ratio of 9:1. New infrastructure, by comparison, was found to be more expensive (ratio <5:1) but offers the highest net benefit of all the tools measured.[15] Still, this result may not mean seawall construction is the best solution. It is, in fact, a salient example of what gets missed in a purely quantitative CBA assessment: new infrastructure may win based on criteria used in the CBA, but as described earlier in this chapter, it fails across a range of other criteria, including environmental justice and co-benefits with climate mitigation. It is perhaps the example of *maladaptation* most often cited.

Methodology matters. Compared with mitigation indicators, such as the tons of carbon emitted, adaptation metrics are far more complex, with nested scales of activity and nuanced overlaps with social protection policies. As shown in the examples above, relying on numerical indicators will likely mean missing dimensions of power sharing, capacity building, and cultural change. Qualitative research methods, including focus groups and structured stakeholder engagement exercises, are more viable for measuring those social dimensions. But to be widely useful, social science methods should integrate with quantitative tools to create standardized indicators. Barriers to the development of an integrated set of shared indicators include scientific uncertainty, spatial diversity, political controversy, and social complexity.[16]

One result of this lack of a coherent set of standards is a reliance on evidence from case studies. But despite a proliferation of these place-based examples, scientists have struggled with methodological challenges that limit rigorous comparative research.[17] Rural examples may not translate well to urban settings. Definitions of common adaptation policy tools may vary in their implementation. And even promising results in a given location may be difficult to attribute to a particular policy intervention. As more adaptation policies are adopted and more rigorous assessments are conducted, this may improve.

Until then, and given scarce resources to address a massive problem like adapting to climate change, experts agree that governments should begin with broadly popular measures, also called *no-regrets policies*. These are neither command-and-control nor market-based strategies per se, but rather are investments over time in adaptive capacity. They tend to be incremental rather than transformational but may represent important foundational opportunities. Two key elements of this kind of adaptive capacity building are policy integration and mainstreaming.

Mapping out different policy responsibilities for various layers of government—as was done earlier in this chapter—is analytically useful, but equally important is identifying the connections among them, also known as *policy integration*. Horizontal integration happens when countries can identify cross-sectoral environmental goals, such as water management and wildlife protection, and create interlocking ministries that share these goals. Vertical integration happens when nested layers of government, like national bureaucracies and provincial administrations, work together on shared goals, maximizing outcomes through layers of authority. Both kinds of coordination rely on institutional strength and focused flexibility.

Relatedly, perhaps the best no-regrets policy approach is *mainstreaming*. When adaptation policies are developed, labeled as such, and implemented as stand-alone law, evidence suggests they are more likely to fail. Funding weaknesses, political interference, and the difficulty of implementing any measure that exists in a silo all serve to undermine those attempts. Instead, weaving adaptation policy into mainstream development, emergency management, social services, and other sectoral policy bundles, can powerfully embed the work into government activity. Over time, this kind of embeddedness—called *mainstreaming*—creates durability. Even in the face of political opposition in a future administration, mainstreamed adaptation policies are more resistant to being canceled.

Some countries and regions have had success with mainstreaming efforts. In particular, the EU has pioneered this strategy, mostly for decarbonization and other mitigation policies. Scholars attribute the EU's success with mainstreaming to the region's shared sense of risk, high level of political commitment, and robust institutional building blocks. But even the mighty EU has stumbled when it comes to mainstreaming adaptation policies; overwhelmingly, "adaptation action appears to be more driven by events than by a strategy of integration and mainstreaming."[18]

A study of southern African countries found some promising success stories for mainstreaming in a region less well known for its policy-making strength. Researchers found that preexisting policies were often substantively unchanged but renamed to better match climate change priorities; this kind of sleight of hand may seem problematic, but since the existing measures were already integrated across sectors, the renaming strategy reaped many of the advantages of intentional mainstreaming.[19] By creating the institutional structures needed for climate adaptation and labeling them as such, future efforts to develop new and more robust approaches can be more smoothly incorporated into the functioning of government. But these isolated examples that illustrate successful mainstreaming are still in the minority; upscaling the approach to the national and international levels has been largely unsuccessful.[20]

Critics observe that in many places, existing government structures are already responsible for creating durable inequities and inefficiencies, and they question the usefulness of building on these embedded systems. Like any tool, mainstreaming can even lead to new unintended problems, as when development and adaptation are conflated into one category of policy making. After all, investment in social services is a tool to reduce both poverty and vulnerability. But for observers, "the integration of adaptation and development can make adaptation more difficult to distinguish from development."[21] The two are not synonymous, and teasing apart the separate strands presents a practical challenge for investors and financial analysts. This lack of boundaries is further confounded through the notion of climate compatible development, a newer concept that seeks to tightly integrate mitigation, adaptation, and development goals. So mainstreaming is desired, but too much embeddedness is problematic; this is, again, a wicked problem.

In this volume, the challenge of defining success is handled with a graceful sidestep. Rather than label some approaches as more successful than others, we provide an overview of common mechanisms and strategies used for climate adaptation governance. Where possible, we explore the dynamics by identifying best practices, providing case examples, and highlighting barriers to effective implementation. Throughout, we consider the ways global power inequities influence the process.

With these foundational pieces in place, we are now ready to dive into the first substantive chapter. Given the prominence of risk assessment in this subject area, we begin with a close look at climate-driven natural disasters. There is some intuitive flow in play here as well, as long-standing governance structures designed to address DRR were among the first to give way to a broader mission that included climate change adaptation.

2

DISASTER RISK MANAGEMENT

Early Warning, Early Action

WHAT IS DISASTER RISK REDUCTION?

Long before weather patterns were definitively linked to global climate change, natural disasters threatened lives, property, and ecosystems. Historically, droughts and floods were the most dangerous, leading to millions of deaths annually and spiraling long-term impacts, including agricultural collapse and rising food insecurity. Today, due in part to improved medical care and urbanization, overall deaths attributable to all kinds of disasters have dropped significantly to an average of 60,000/year,[1] while absolute costs associated with natural disasters have risen steadily, although not evenly, to an estimated $232 billion in 2019.[2] Figure 2.1 illustrates this pattern.

Scientists now understand that many natural disasters are influenced by climate change in a variety of complex ways. Attribution science—the study of the extent to which climate change influences weather and climate events—is one of the fastest-growing fields in climate research.[3] Scientists in this emerging field have to overcome methodological challenges that include separating attribution from natural variability. Amid these complexities, a scientific consensus has emerged: climate

Cost Of Climate-Related Disasters Soars 150%

Total economic cost of global climate-related disasters over 20-year periods

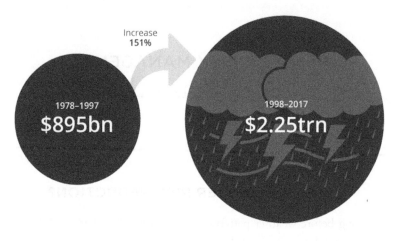

statista ⚡

FIGURE 2.1 Both frequency and cost of natural disasters have increased,
attributable largely to dynamics associated with
anthropogenic climate change.

Source: Statista; United Nations Office for Disaster Risk Reduction.

change contributes to both the likelihood and the magnitude of
many disasters. It's worth noting that earthquakes, the deadliest
type of natural disaster in recent years, are generally not attributable to climate change; still, building risk management systems
for climate-induced disasters may have spillover benefits for earth-
quake response.

As described in chapter 1, hazards such as floods, wildfires,
volcanoes, and earthquakes are largely not preventable, at least
not in a proximate sense. But loss of life, irreversible damage to
ecosystems, and destruction of property are all elements of exposure that can be forecast and then mitigated through focused
DRR. Understood as the interaction between physical processes

and the relative vulnerability of exposed elements, DRR "works at the intersection between understanding risk and risk impact [to] reduce disaster loss and prevent the emergence of new risk."[4]

Not surprisingly, catastrophic disasters are most damaging in densely populated urban areas and in communities already saddled with social vulnerabilities and capacity weaknesses. But of course, even wealthy, high-capacity communities have been clobbered by natural disasters. Globalization has created durable networks around trade, communications, and finance, so a disaster that occurs in one location can have dramatic impacts on distant landscapes. It is this very complexity that bedevils our ability to respond comprehensively. Perhaps the UN's annual report on disaster risk management summarizes the challenge best: "Systemic, cascading risk is the underlying feature of today's risk landscape."[5]

GOVERNANCE OF DISASTER RISK REDUCTION

Leaders of early efforts to manage disasters treated them as external hazards that could be managed by stronger construction methods or better storm shelters, an approach that leads inexorably to a technocratic framing. As described by one observer: "The identification of disasters as purely physical occurrences . . . that affect people who have the misfortune to be simply in the wrong place at the wrong time gave rise to a preoccupation with technological solutions for the protection of infrastructure and exposed populations."[6] This incomplete understanding of how hazards work ignores the powerful role of structural vulnerability in determining which individuals and cities will be hardest hit by incoming hazards, which will be able to recover, and what steps will best reduce undesired outcomes.

Global Governance

Fortunately, our collective understanding of the intersection between hazard and risk has evolved. Global governance for DRR today is largely organized around a robust UN-led institutional framework. In 2000, the UN International Strategy for Disaster Reduction (UNISDR) was established, the precursor to today's UN Office for Disaster Risk Reduction (UNDRR). Like other UN program offices, UNDRR explicitly recognizes state sovereignty, deferring to national governments. Participation is voluntary. The Hyogo Framework for Action, established in 2005, was the primary instrument for organizing national-level disaster management efforts until 2015, when the Sendai Framework for Disaster Risk Reduction (Sendai Framework) was established alongside the PA and the Sustainable Development Goals (SDGs) as part of the broader 2030 Agenda for Sustainable Development (figure 2.2). With four interlocking priorities, the Sendai Framework encompasses scientific, humanitarian, financial, and social dimensions of risk management. Progress toward seven global targets is measured by thirty-eight indicators, all tracked through the Sendai Framework Monitor, an online tool where member states can report data. By 2019, the vast majority of countries (130) were actively using this voluntary system, suggesting improved accountability for national efforts.

As part of its mission, the Sendai Framework includes targeted outreach to regional and national disaster management authorities and the private sector. Regional disaster planning has been taken up in Africa, Europe, and other areas, and the resulting cross-boundary disaster management has yielded more efficient recovery outcomes. Those regional institutions are then well positioned to support better national-level planning by their members, leading to improved policy coherence. In these ways,

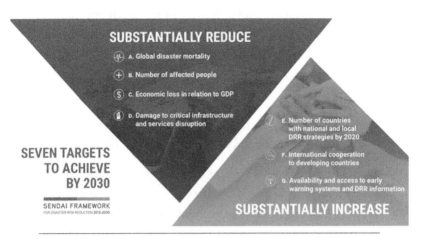

FIGURE 2.2 The Sendai Framework is focused on reducing risk from natural disasters.

Source: United Nations Office for Disaster Risk Reduction.

the value of global DRR comes into focus: it builds capacity in states through technical support and helps to standardize data collection around the world.

Still, and despite the scientific consensus that climate change and natural disasters are connected, policy making in those two realms is not well integrated, and complex barriers persist. For example, DRR practitioners are likely to orient their work in the present or recent past as they seek to manage an incoming disaster or to address recovery needs of hard-hit communities. Climate adaptation planners, by contrast, focus most of their attention on the future, anticipating how impacts will affect particular locations. But we can envision a stronger convergence between the two that would yield a range of benefits for both fields: among them, linked systems could foster stronger assessments of vulnerability, more robust use of data, better connections between global patterns and local realities, and improved efficiency. Global leaders in the DRR regime now explicitly recognize these possibilities. Policy alignment and strengthened partnerships were two of the top objectives for the UNDRR in 2019.

National, Subnational, and Private-Sector Governance

Domestic governments tend to organize their attention to this issue through a dedicated ministry or other high-level agency charged with managing disaster events; in the United States, for example, the Federal Emergency Management Agency (FEMA) has historically been the lead institution for disaster management. As might be expected, many of these public agencies face pervasive funding shortages, and some are hobbled by insufficient predictive scientific expertise. They are also subject to political capture, as in the case of FEMA under President Donald Trump: its *2019 National Preparedness Report* never mentioned climate change or sea-level rise.[7]

Subnational governments have also taken on many responsibilities for DRR, often under the institutional umbrella provided by the Sendai Framework and national strategic planning. Major metropolitan areas have been particularly aggressive in their efforts to plan ahead for recurring disasters (see chapter 4 for details on urban planning). In many cities, authorities look to predictive data to build cost estimates and then set aside emergency funding through public budgeting processes. They may develop risk maps to guide development and land use. They may focus on retrofitting infrastructure and building new structures that can better withstand predicted weather patterns. They may even pursue broader social reforms that address entrenched patterns of vulnerability. But these smaller governmental units face substantial procedural obstacles that stem from the limits on their jurisdictional authority. For example, in the United States, the tristate area of New York/New Jersey/Connecticut has 3,700 miles of coastline and twenty-three million residents who regularly face storm surge and hurricane damage. The region has no dedicated budget for coordinated adaptation planning and instead leaves DRR to the individual states. Layers of competing jurisdictional authorities magnify administrative barriers to comprehensive DRR (figure 2.3).

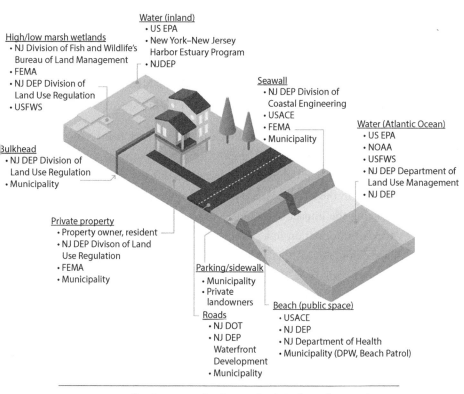

FIGURE 2.3 Overlapping, redundant, and uncoordinated ownership jurisdictions are a challenge for disaster risk reduction efforts.

Source: Image adapted courtesy of Regional Plan Association.

Interest in disaster risk extends to the private sector, where investors at all scales evaluate relative risk before allocating funds. The Global Risk Assessment Framework was launched in 2019 to facilitate interdisciplinary research and forge a pathway for communicating findings to the private sector, with the hope that such data will "result in better risk-informed strategies, plans and investments."[8] Insurance, usually offered through the private sector but increasingly subsidized by governments, plays a critical role in disaster planning and recovery. See chapter 6 for a deeper consideration of the role of insurance in managing risk and chapter 8 for more on private-sector investment.

PLANNING: EARLY WARNING AND EARLY ACTION

At all levels of government, much of the disaster management work is organized around improving Early Warning Systems (EWS). First highlighted through a series of early warning conferences in the late 1990s, EWSs "are complex processes aimed at reducing the impact of natural hazards by providing timely and relevant information in a systematic way."[9] Copious case studies have demonstrated that effective EWSs can reduce both casualties and costs from disasters. Synthesizing findings from those cases, the Global Commission on Adaptation found in 2019 that predisaster investment in improved EWSs resulted in cost-benefit ratios of at least 10:1. With just twenty-four hours of warning for an incoming disaster, the ensuing damage can be reduced by 30 percent.[10]

Given this compelling evidence for the effectiveness of the tool, decision makers are keen to build capacity along the four axes of an EWS: risk knowledge, monitoring, dissemination, and response. *Risk knowledge* comes from scientific data, including satellite information, weather station information, and modeling. That knowledge must also include sufficient historical context—baseline data—so that experts are able to identify emerging threats. During a monsoon, for example, risk assessors have to know whether recent rainfall is seasonally appropriate or an emergent flood risk. To adequately measure risk, those assessors also need robust data on the exposure and vulnerability of local populations. *Monitoring* should be iterative and scaled so that preidentified trigger points are quickly noticed. Real-time data collection relies on a range of technological forecasting tools managed by scientists; in some locations, it also includes the uploading of observational data by citizens with mobile phones. Once a proximate risk is identified, *dissemination* becomes critical. Communication tools may include television broadcasts, radio alerts, messaging through mobile phone apps, and audible sirens. In many locations, planning for effective dissemination means contending with multiple languages and reaching people who may be illiterate, disabled, or geographically disconnected, an objective embedded in the notion of "human-centered" EWSs. Effective outreach must include details about the desired *response*. Should alerted residents evacuate to a nearby shelter? Has that shelter been equipped to provide food and medical care? Do residents believe they will be safe if they leave their homes? Should they take other immediate action to reduce their exposure? Answering these questions in the chaos of a disaster is unworkable and should be done ahead of time as part of a comprehensive EWS.

Barriers to effective EWSs abound. Capacity weaknesses, lack of scientific expertise, institutional disorganization, jurisdictional uncertainty, and absence of necessary technology can cripple governments, making it difficult to mount the complex components of an EWS. Without a cooperative and widely disseminated media platform, alerts may not reach their intended audience. And without sufficient predisaster communications, members of the public may not adequately understand the significance of government-issued warnings. Progress is uneven. In 2018, researchers found that out of 114 low-income countries, approximately half had implemented human-centered EWSs. Many of those countries also reported a subsequent reduction in the number of people affected by or killed during disasters.[11]

Global and regional institutions have now developed technical guides that can help decision makers navigate trade-offs in the establishment of EWSs, and case studies from around the world also contribute to a better understanding of best practices. Four themes emerge. First, integrated planning yields the strongest results. Mixing risk assessments with management plans and then linking those to hazard mitigation efforts can create durable and mainstreamed institutional foundations. Online support tools such as the EU's Climate ADAPT portal offer templates for navigating this. Second, robust and human-centered communication tools should be the centerpiece of any EWS. Examples across rural landscapes in the developing world highlight creative ways practitioners are making data available to households in remote locations through, for example, user-friendly apps on ubiquitous mobile phones. Third, attention to timescale matters. Signs of a long-term drought pattern, for example, should lead to a different decision pathway than would evidence of an incoming hurricane. Focusing on immediate and short-term threats—CRM, as described in chapter 1—may not lead to long-term adaptive solutions, but it can reduce damage in the short term and build longer-term capacity.[12] Fourth, social safety nets play an essential role in the execution of EWSs. Disparities in access to services occur both within and between countries, vividly highlighting how subsets of a single population can experience different outcomes from the same disaster. See chapter 8 for a deeper exploration of climate adaptation in an unequal world.

CASE STUDY: HURRICANE KATRINA

Hurricane Katrina (2005) captured international attention when images of New Orleans residents stranded on rooftops blanketed the news cycle. Levees designed to protect the low-lying city from storm surge failed, submerging sections of the city. An estimated one million residents were displaced, many of them landing in one of 707 Red Cross emergency shelters established across neighboring states. Low-income Black residents were found to be

BOX FIGURE 2.1 Locator map of New Orleans, LA.

significantly less likely to heed evacuation warnings—probably a result of deep-seated distrust for government officials in a part of the country where institutionalized racism has deep roots—contributing to a disproportionately high death rate for that demographic. After the hurricane subsided, residents who owned their homes were more likely to return to the city and participate in recovery efforts; low-income Black residents were significantly underrepresented in that group, and even today, the city has not regained its prestorm racial balance. The experience of Katrina highlights how preexisting economic vulnerability and relative social trust can affect outcomes in a disaster.*

*James R. Elliott and Jeremy Pais, "Race, Class, and Hurricane Katrina: Social Differences in Human Responses to Disaster," *Social Science Research* 35, no. 2 (2006): 295–321, doi:10.1016/j.ssresearch.2006.02.003.

CLIMATE SERVICES

Without question, the availability of updated climate data is central for effective DRR at all scales. And, yet, research suggests that use of climate information is not widely integrated into decision-making, suggesting a breakdown in the delivery and/or usefulness of that data.[13] To solve this problem, practitioners urge targeted communication to overcome such cognitive processes as heuristics, discounting, and uncertainty bias that interfere with the way scientific information resonates with individuals. After all, data spurs action only if they are communicated effectively. *Climate services*, defined as "scientifically based information and products that enhance users' knowledge and understanding about the impacts of climate on their decisions and actions," captures both sides of this coin: the production of knowledge and the effective communication of that knowledge.[14] Increasingly, EWSs are forecast based, meaning they incorporate not just data about weather events that have already occurred but also predictions for near-term trends that pose risks.

Scientific knowledge is produced in an uncoordinated manner, in both the public and the private sectors, through a wide variety of funding mechanisms and administrative arrangements. As described in chapter 1, the IPCC offers the highest level of synthesized peer-reviewed climate data in the world. But IPCC data are largely global and long-term and, as a result of the scale mismatches across these two axes, are not well suited for localized, shorter-term DRR efforts like EWSs. Further, given the historical silos that have separated DRR from climate change within the UN system, even relevant scientific data about disasters may not be immediately applied to adaptation settings. National governments might seem like a more useful place to house scientific data, but they vary widely in their scientific capacity: some well-funded countries support robust climate modeling and predictive scientific data collection, while others have far less financial and human capital. At the local level, where so much adaptation work happens, communities may have generations of place-based knowledge but a scarcity of verifiable scientific data. Social scientists suggest pairing the two kinds of knowledge through a process of co-production so that indigenous ways of knowing are integrated with western quantitative models (see chapter 8), an important but difficult endeavor.

Consistent historical records indicating the frequency of natural events at a localized or regional scale are essential for tracking deviations from the norm. Especially when trying to assess possible manifestations of compound events, planners need climate models that reflect both mean state and natural variability.[15] Countries with recent histories of civil unrest are at a particular

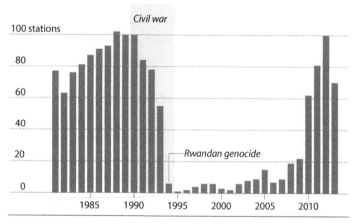

FIGURE 2.4 The Rwandan genocide disabled the country and resulted in incomplete weather station data for nearly two decades.

Source: International Research Institute for Climate and Society.

disadvantage, as they may not have consistent weather station data (figure 2.4).

Even in wealthier and relatively stable countries, where decision support tools have been built specifically to provide local officials with up-to-date climate information, efforts to provide climate services have fallen short. The Adaptation Wizard, developed in the United Kingdom (UK) and made available in 2004, was one of the first such comprehensive tools, and yet researchers in 2018 found that it is not highly relevant for adaptation planning, in part because it is incompatible with the needs of small communities there. Local decision makers report wanting hands-on support for specific projects rather than the broad guidance offered through the tool.[16] Other decision support tools may fare better. Those that provide local-level support with minimal preconditions or expectations for knowledge appear to be the most useful. The preliminary lesson is that predictive tools must better connect with local needs and capacities.

But how can we think about forging these connections in a context of information asymmetries? Sometimes decision makers simply don't know what they don't know; therefore, "a key challenge for developing adaptation support tools is to find a balance between simply meeting target group expectations and enlightening them at the same time."[17] Further complicating the movement of scientific knowledge from the laboratory to the field are disparities in internet access and mobile phone technology. Globally, 61 percent of people were in possession of mobile technology in 2020, but the numbers range wildly, from 82 percent in the UK to only 15 percent in Nigeria.[18]

Some countries have addressed these hurdles through targeted community outreach that prioritizes face-to-face communication of scientific information. Farmer-to-farmer extension services have been shown to be especially useful for local decision-making, and farmer field schools (FFSs) have emerged throughout the developing world to meet this need. Those FFSs work through existing cultural hubs like churches to educate farmers and empower them to build improved climate knowledge within their own communities. This kind of peer learning is a particularly effective approach, as people everywhere tend to trust and mimic the behavior of their close social contacts. Still, research has shown that nothing can quite match the power of individual experience to heighten awareness: "Farmers will be more willing to adopt adaptation strategy if they individually perceive climate change damage."[19]

CONCLUSION

Reducing risk from disasters is one prong of a comprehensive climate change adaptation approach. Doing so means being attentive to the integration of global, national, and local conditions, using robust EWSs in the process of matching the Sendai Framework's overarching priorities with the experiences and capacities of communities where the inevitable disaster will arrive. We now turn our attention to another, related dimension of the adaptation challenge: the role of the built environment in mitigating risk.

3

THE BUILT ENVIRONMENT

Infrastructure and Nature-Based Solutions

NY analysis of climate change impacts invariably
includes attention to infrastructure, as many cities
have discovered that long-standing construction practices are no longer durable in a climate-altered world. Energy
grids, transportation systems, and storm barriers may be unable
to withstand higher temperatures and growing storm intensities.
Further, the use of some common materials—most notably,
concrete—is now understood to generate dangerous GHG
emissions. As a result, many countries have begun to explore
alternative materials and supply chains as they aspire to climate change mitigation targets. Even infrastructure development that is intended to be resilient can contribute to patterns
of social disparity and path dependency. For all of these reasons,
the role of infrastructure in a climate change adaptation strategy
is contested.

The World Economic Forum's *Global Risks Report 2019*
suggested that 90 percent of the world's coastal areas face risks
to human life, property, transportation networks, water systems,
and ports.[1] Rising seas and compound risks associated with
storm surge have created persistent flood zones. Estimates suggest that by 2050, flooding along coastlines around the world

will cost $1 trillion annually.[2] For low-lying coastal communities faced with the threat of rising seas and flooding from increasingly intense storm surge, building a protective physical barrier is intuitive. Indeed, many cities have responded by building levees and seawalls, both examples of "hard" or "gray" infrastructure. In the United States, for example, the vast majority of national efforts to address flooding risk by FEMA and the U.S. Army Corps of Engineers have involved hard (or gray) infrastructure. Billions of dollars have already been spent, and, yet, only 14 percent of the coastline has been protected.[3] Flooding persists.

Now, as intensified weather conditions test the viability of those hard coastal defenses and as costs continue to climb, many communities have begun to integrate less expensive nature-based solutions (NbS). Those projects rely on existing or improved ecosystem features to provide societal benefits. For the built environment, we might think about adaptation options along a continuum, with traditional hard infrastructure at one extreme and NbS at the other (figure 3.1). In the middle section of that spectrum lies a variety of options that practitioners call green or resilient infrastructure, suggesting modifications to traditional

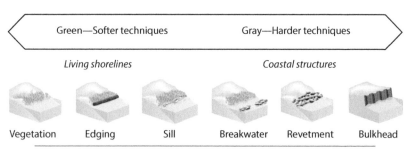

FIGURE 3.1 A variety of options exists for protecting coastal areas, ranging from hard infrastructure to natural vegetation.

Source: Image adapted from NOAA.

construction methods that take climate change into account. At every stage, deciding which approach to pursue will involve trade-offs in expense, uncertainty, and protection.

TRADITIONAL HARD INFRASTRUCTURE

Since the earliest days of recorded history, human societies have met a variety of social challenges with infrastructure solutions. Our cities are built on webs of infrastructure that allow for transportation, energy distribution, water storage, and communications. For many, the very notion of development is synonymous with the creation of new, powerful infrastructure. For example, urban areas everywhere struggle to manage storm-water runoff. Most have tackled this challenge by constructing roadside drains and sewers to carry unwanted water away from impervious surfaces. Those urban water-management features provide a relatively efficient and well-understood solution to a recurrent natural resource management problem. Indeed, infrastructure development has been the favorite not only of governments but also of investors. It is well understood. The product is tangible. It is durable. When integrated with nonstructural measures like policy to guide siting and funding, it contributes to perceptions of modernization and progress.

These engineered, technical solutions to disaster risk comprise the majority of publicly funded activities. This is true for most countries regardless of economic capacity.[4] To take one well-known example, Bangladesh, one of the countries most vulnerable to coastal disasters, relies heavily on hard interventions. From 2009 to 2016, its national Climate Change Trust approved 329 climate adaptation projects, with 88 percent of them (291) taking the form of hard infrastructure.[5]

Yet, despite its popularity, hard infrastructure is problematic in the context of climate change. It tends to rely heavily on fossil fuels for both construction and maintenance, making it prone to maladaptation. It is particularly vulnerable to the urban heat island effect—wherein dense collections of buildings create pockets of warmer temperatures—and performs poorly in extreme weather, creating instability and generating new risks for localities.[6] When the Organization for Economic Cooperation and Development (OECD) modeled flood risk in Paris, it found that as much as 55 percent of the direct damage would be borne by the infrastructure sector, creating high recovery costs. That infrastructural damage, including harm to transportation corridors and the energy grid, would then drive up to 85 percent of business losses in the city; indeed, the infrastructure itself could quickly become a source of disruption.[7] This projected situation played out in real time in New York City in 2012 when Hurricane Sandy knocked out transformers and flooded rail lines. That infrastructure damage led to widespread blackouts and created months of mobility delays.

Hard infrastructure is also uniquely positioned to create path dependency simply because it brings a level of permanence not associated with other measures. Artificial snow and the infrastructure to deploy it are illustrative. In ski resorts around the world, the ability to deliver reliably good snow cover through industrial-scale snowmaking machines has given rise to housing development, tourism growth, and localized economic dependence on winter recreation. One researcher notes that these patterns have become rote in many locations: "Winter tourism is an industrial sector that has undergone little systematic or science-based planning . . . [it follows the] same sequence of infrastructural development: artificial snow, water-holding reservoirs, ski-run enlargement, lifts, roads and housing."[8] When climate change disrupts historical precipitation patterns, the snowmaking machines become increasingly central to local resilience. The snowmaking infrastructure is, in some important sense, wagging the dog.

At the heart of the concern over building infrastructure lie the inaccuracies caused by using historical weather patterns to predict future climate. Since the past no longer provides a complete picture of what's possible in the future, cost estimates and projections of effectiveness are likely to be wildly inaccurate. Existing planning modes too often lead to this mistake, resulting in "built systems [that] only cater to a specific subset of possible futures."[9] Significant work has gone into understanding how infrastructure can fail in extreme weather and how better engineering outcomes can be achieved, but critics note that upgrading infrastructure will not be sufficient. Reformation of the U.S. power system, for example, may result in an upgraded grid, but that new climate-smart infrastructure will succeed only if it is matched with appropriate public policies and underwritten with civic support.[10] Attention to engineered solutions must be combined with social, political, and cultural understanding.

Yet, commitment to hard infrastructure runs deep. In some countries, this may be attributable to cultural values that favor modern engineering solutions; for example, researchers report that in Vietnam, where social norms tend to prioritize technology, robust adaptation planning still relies overwhelmingly on "hard" solutions for DRR.[11] In other locations, researchers observe that countries with siloed institutions are more likely to favor infrastructure solutions and that those approaches in turn create a form of path dependency, solidifying both the isolation of each silo and the mandate to build better structures to meet evolving risk.

Politics can also drive preferences. In the wake of Australia's massive wildfires, Prime Minister Scott Morrison announced his government's intent to focus on climate adaptation and resilience instead of mitigation. He asserted that "building dams is climate action now."[12] Debate around Australia's climate policy is highly politicized, so it may be politically strategic for its leaders to reframe climate change as inevitable and impossible to reverse while claiming old patterns of development—hardened water storage—are now adaptive. Other leaders have similarly repackaged and green-washed infrastructure development as a solution to climate adaptation needs.

RESILIENT INFRASTRUCTURE

As concerns about the climatic effects of hard infrastructure have increased, engineers have begun to forgo traditionally built structures in favor of ones that use more resilient materials and forms. These newer approaches include a suite of departures from hard infrastructure construction: changes in planning processes, new decision methodology, the use of different materials, and development of innovative designs. In its fullest expression, resilient infrastructure refers to projects that are part of "a network of green features that are interconnected and therefore bring added benefits and are more resilient than if they were isolated."[13]

Green roofs are an example. Long seen as a climate mitigation strategy—providing new sources for carbon sinks—the installation of green roofs is now widely embraced as having adaptive advantages too. They can help keep buildings cooler, thereby reducing the need for maladaptive air-conditioning.[14] They also provide a range of attractive amenities for residents, including venues for recreation, urban gardening, and bird-watching. These pursuits may seem trivial, but remember that enhancing the quality of life in dense urban settings is consistent with efforts to reduce vulnerability and build adaptive capacity. An iconic example of one low-cost measure yielding both mitigation and adaptation benefits—called co-benefits—green roofs are also perhaps an ideal example of a no-regrets project.

Another example is *blue-green infrastructure* (BGI), which refers to a set of features that work together specifically to address urban storm-water management. Such systems may include rain gardens, constructed wetlands, retention basins, and bioswales. Where gray infrastructure projects tend to serve a single function, BGI has been shown to provide multiple ecosystem benefits alongside flood reduction and improved water quality.[15] In some instances, existing gray infrastructure can be retrofitted with new climate-smart tools, transforming maladaptive structures into BGI systems. Overall, the costs associated with this kind of retrofitting, combined with those of periodic maintenance, can be substantially less than the costs of brand-new construction, and the results can be profound. In one study, researchers in Sweden found that retrofitted green storm-water systems in a high-risk neighborhood led to a 90 percent reduction in flood magnitude.[16]

Public support for BGI is high. Even when residents do not understand the climate change dimensions in play, they tend to view new green infrastructure projects as having personal advantages for their lives, including increased recreation opportunities and reduced heat in urban settings. These preferences also translate into a willingness to pay for such features, suggesting financial support might exist for expanded use of these tools.[17] Policy makers would do well to remember that people's support for this type of new construction is less about how the new features can contribute to long-term climate adaptation and more about benefits in the near term for themselves and their families. Resilient infrastructure offers an effective tool for reducing exposure, mitigating risk, and building public enthusiasm for a new development paradigm.

A note of caution is warranted here regarding the connections between updated resilient infrastructure and the perils of gentrification. Certainly, climate adaptation measures do not benefit all residents equally; indeed, this theme resonates throughout this volume. Urban planners are only beginning to understand the dynamics in play here, but early research suggests that "the risks prioritized by socially vulnerable groups (displacement, physical insecurity) are deprioritized in the name of addressing identified climate risks through green infrastructure."[18] If vulnerable groups face increased risk of displacement as wealthier communities thrive in newly constructed resilient neighborhoods, this is the very definition of gentrification.[19] Whether the displacement is intentional or incidental, this outcome suggests an insidious note of maladaptation in the heart of an otherwise promising development pathway.

NATURE-BASED SOLUTIONS

Some features described earlier as resilient might also be categorized under the broader heading of NbS, defined as "actions to protect, sustainably manage and restore natural or modified ecosystems that address societal challenges effectively and adaptively, simultaneously providing human well-being and biodiversity benefits."[20] The notion is rooted in our understanding of *ecosystem services*, the benefits delivered to human societies by healthy ecosystems, usually free of charge. For example, healthy mangrove forests can serve as essential flood mitigation tools, absorbing seawater and buffering coastal communities from hazard. Mangroves grow naturally along coastlines in tropical and subtropical regions—including much of Latin America, Oceania, Africa, and Asia—but are increasingly threatened by coastal development and other human activities. Increasing the extent and improving the health of those mangroves may mean protecting existing species and/or planting new ones; these NbS can help to maximize the value of the ecosystem services provided.

Digging deeper into the ecosystem services framework, three gradations of NbS can be plotted along a continuum. First, communities can protect and use existing natural systems. In the mangrove example, this might mean formally designating those valuable forests as protected areas to ensure their survival. Second, societies might explore new ways to manage existing ecosystem features to enhance benefits. Degraded ecosystems can be restored through focused management; the mangrove forests might be woven into new agroforestry systems to maximize

CASE STUDY: COSTA RICA

When Costa Rica's biodiverse rain forests began to disappear in the 1970s and '80s, the loss was attributed to a burgeoning banana trade and the growth of agricultural practices that relied on livestock. Other Latin American countries faced similar pressures at that time, but Costa Rica's rate of deforestation was the highest in the region; the nation's inland forests were decimated, and an estimated 40 percent of coastal mangrove coverage was lost. Concentrated policy efforts began in the 1990s to reverse the trend and included the innovative Payment for Ecosystems Services program along with improved enforcement to prevent illegal logging. Restoring coastal mangroves through an NbS approach was essential for reducing risk from frequent storm surges. Pilot restoration programs began in the vast Terraba Sierpe

BOX FIGURE 3.1 Locator map of Costa Rica.

National Wetland. As invasive ferns were removed, the mangroves were reestablished, and local communities soon discovered the healthier ecosystems supported a resurgence of mud cockles, a locally valued mollusk. These promising results highlight how using NbS to restore degraded landscapes can contribute to both the economic viability of the community and the realization of climate change adaptation goals.

their impact, for example. Third, land managers might create a new ecosystem where it never existed or had previously been eradicated. Planting new mangroves along coastlines might increase flood protection. Many NbS projects combine these three elements for maximum benefit. Taken together, investments in planting and maintaining mangrove forests along high-risk coastlines have been shown to contribute $80 billion a year in avoided losses from flooding, thereby protecting eighteen million people around the world. Nonmarket benefits of the practice include enhanced fisheries, healthier forests, water filtration, and the creation of new recreation opportunities, with total benefits estimated to be ten times the cost.[21]

Practitioners and scientists have hailed NbS as an interdisciplinary solution for interlocking global challenges. With benefits for climate mitigation, climate adaptation, biodiversity recovery, and sustainable development, the strategy would appear to be low on trade-offs and high on results (figure 3.2).

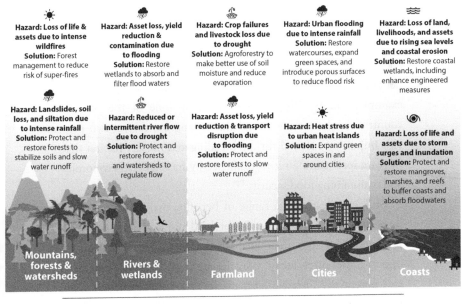

FIGURE 3.2 NbS can function alone or in concert with hard infrastructure to maximize ecosystem services and build resilience.

Source: Image adapted from Global Center on Adaptation.

Avoiding expensive inputs and high-tech engineering requirements, NbS can deliver powerful benefits for a fraction of the
cost of more traditional "hard" solutions.[22] Once planted, those
mangroves are likely to be more resistant to political whims
than flimsier policies that rely on a steady stream of funding and
political support. In this way, NbS become systemic solutions.[23]
Even the private sector is starting to embrace NbS, as they "are
increasingly being viewed not only as a way to reconcile economic development with the stewardship of ecosystems, but
also as a means to diversity and transform business and enable
sustainable development."[24]

Often interwoven with gray infrastructure development or
other tools, NbS may also reflect indigenous relationships between
humans and the natural world, creating an important bridge
between modern engineering practices and localized, customary
knowledge. Indeed, the affordability of NbS means some of
the most promising early experimentation is coming from the
developing world. Benefits include not only restored ecosystems
and enhanced adaptive capacity but also powerful gains in human
development and poverty alleviation. Returning to the mangrove
example, communities in Mozambique—one of the poorest
countries in the world—found that "mangrove forests also provide important fish and shrimp breeding and nursery habitat that
increase local communities' resilience by fueling local economies
and maintaining food security."[25]

Still, and despite the range of benefits, many countries face
powerful competing demands for land use and have failed to
integrate NbS into their adaptation planning.[26] These missed
opportunities often mean countries are choosing more expensive
and potentially maladaptive hard (or gray) infrastructure solutions instead. As successful examples of NbS spread, perhaps
more communities will prioritize those approaches.

DECISION-MAKING PROCESSES

So far, the most promising approach for infrastructure development seems to be a combination of gray and green tools. The challenge is figuring out the best ways to combine those strategies to meet the contextual needs of risk and capacity at the local level. Maximizing these synergies means focusing on the effectiveness of the tools, development of governance pathways, and analysis of trade-offs. But for infrastructure, like many of the adaptation tools profiled in this volume, measuring effectiveness is fraught with methodological challenges. Robust CBA that includes various scenario pathways and the potential for compound risk can help sift through the choices. For example, a comprehensive analysis of existing and planned adaptation measures along the U.S. Gulf Coast concluded that the NbS were the most cost-effective. Home protection methods like sandbags and dikes also had high success rates but were very expensive.[27]

The decision-making process provides another critical opportunity to integrate lessons learned from successful adaptation cases. Rather than following the traditional model—a top-down, technocratic, infrastructure-inclined pathway—adaptive governance offers a way toward more bottom-up, co-produced, and climate-smart solutions. Given the range of resources and ecosystems impacted by NbS, "to be successful, governance of NbS requires (and indeed enables) active cooperation and coordinated action between stakeholders whose priorities, interest, or values may not align, or may even conflict."[28] This mandate presents tantalizing opportunities for a synergistic approach with both process and outcome benefits.

But how can stakeholders—including the decision makers themselves—sort through the trade-offs to determine which kinds of infrastructure belong where? One answer comes from

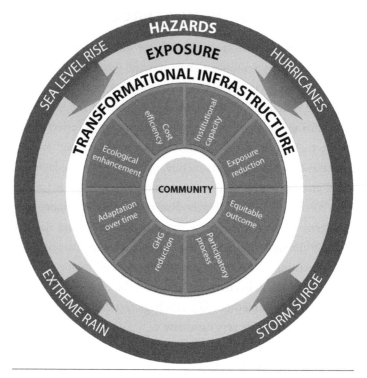

FIGURE 3.3 Determining which kind of infrastructure will be most resilient can be approached using the Adaptive Gradients criteria model.

Source: Image adapted courtesy of Yaser Abunnasr.

the ambitious Adaptive Gradients project (figure 3.3); in 2018, it created a framework that enables researchers to assess the relative success of infrastructure projects by ranking them across eight gradients: exposure reduction, cost efficiency, institutional capacity, ecological enhancement, adaptation over time, GHG reduction, participatory decision-making processes, and social benefits.[29] Updated indicators and standardized metrics could help more communities navigate these complex trade-offs.

CONCLUSION

Many of the infrastructure and NbS examples provided in this chapter come from urban settings. In the next chapter, we move from this overview of the built environment to a closer look at those same cities, where we explore the unique climate adaptation challenges faced by dense human settlements.

4

URBAN PLANNING FOR
CLIMATE ADAPTATION

URBANIZATION AND DEVELOPMENT

One of the defining features of global development has been a steady march toward urbanization. As cities become centers for industry, rural residents relocate in search of improved economic opportunities, better education, and more reliable access to other public services. Today, urban areas are home to more than half of the global population. More than four hundred cities have a population that exceeds one million. The trend has manifested unevenly, with some parts of the world more urbanized than others. Fully 73 percent of European residents live in urban areas, while far lower proportions are urban across much of Africa (43 percent) and Asia (50 percent).[1] Megacities—defined as those urban centers with more than ten million residents—now number thirty-five, with the majority in India and China (figure 4.1).

Cities have long been seen as innovation hubs, as they tend to be organized around higher education, technology, and entrepreneurship. Overall, public services, including health care and transportation, are more widely available in urban areas than in rural ones. Cities are engines of growth and centers for global trade. But within every city, wealth and privilege vary widely.

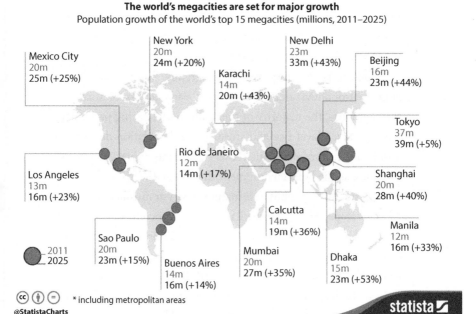

The world's megacities are set for major growth
Population growth of the world's top 15 megacities (millions, 2011–2025)

Mexico City
20m
25m (+25%)

New York
20m
24m (+20%)

Karachi
14m
20m (+43%)

New Delhi
23m
33m (+43%)

Beijing
16m
23m (+44%)

Tokyo
37m
39m (+5%)

Los Angeles
13m
16m (+23%)

Rio de Janeiro
12m
14m (+17%)

Shanghai
20m
28m (+40%)

2011
2025

Sao Paulo
20m
23m (+15%)

Calcutta
14m
19m (+36%)

Manila
12m
16m (+33%)

Buenos Aires
14m
16m (+14%)

Mumbai
20m
27m (+35%)

Dhaka
15m
23m (+53%)

* including metropolitan areas

@StatistaCharts

statista

FIGURE 4.1 Global megacities are dense urban settings
with more than ten million residents.

Source: Statista; UN Population Division, World Economic Forum

Many of the world's largest cities are also home to profound poverty, with an estimated 880 million people living in informal urban settlements.[2] Characterized by insecure land tenure (see chapter 5), poorly constructed shelters, high rates of disease, and often high rates of crime, these urban slums are also highly vulnerable to natural hazards. In some cities, this problem is a relic of colonial histories. For example, across much of the African continent, colonial powers left residents across the countryside with weak land tenure systems, thereby contributing to urbanization pressures; the result has been "the growth of extensive informal settlements and slums in almost all cities, towns, and mining centers in the [sub-Saharan African] region."[3] Those settlements are highly exposed to hazard, extremely vulnerable, and ill-equipped to rebound from natural disasters.

Exposure is also amplified by geography. Since most of our large metropolitan areas grew out of early access to markets through ports, the majority of them are coastal, facing particular risks from sea-level rise and storm surge. Population density— a defining feature of urban centers—presents more challenges for cities, as the built environment that houses, feeds, and transports residents can obliterate natural ecosystem function from the landscape. Heat absorbed into those concrete jungles gets stuck there, contributing to higher temperatures in cities than in surrounding towns, a phenomenon known as the *urban heat island*. Deadly heat waves have claimed thousands of lives; efforts to beat the heat compel urban residents to increase their use of air-conditioning, thereby emitting more GHGs and unwittingly participating in autonomous maladaptation.

Perhaps these challenges can be overcome by the talented and well-resourced individuals who live in urban areas. After all, we know that cities on average have better-educated residents, a stronger tax base, and more access to investment. They tend to have more robust government structures in place and be more firmly connected to policy-making centers. Many cities are also hubs for "municipal voluntarism," characterized by voluntary, spontaneous actions to improve living conditions that occur outside of coordinated policy efforts.[4] Despite these attributes, the vast majority of cities currently have no explicit adaptation initiatives underway.[5] Many address disaster risk but not adaptation; in other words, they consider short-term risks from natural disasters but aren't planning long-term comprehensive climate change adaptation.

It's worth noting that these data may well be incomplete. A lack of progress reporting is not synonymous with a lack of action. Those cities that have successfully mainstreamed climate adaptation—as recommended—may be particularly difficult to assess, since adaptation policies are embedded in other issue areas and hard to identify. Further complicating this assessment, municipalities in some countries face severe limitations to their authority; they are not considered change agents, and, instead, regional, state, or national policy officials take the lead. In those instances, policy changes that impact urban adaptation may be credited to national decision makers rather than local leadership. The pattern holds even in highly decentralized countries; counterintuitively, many such cities lack autonomy. They contend with inadequate funding and have limited authority to develop independent initiatives that might contribute to their own capacity for climate adaptation.

URBAN PLANNING: ZONING, LAND USE, AND BUILDING CODES

In many countries, cities' most salient leverage point is urban planning. Through zoning and related measures, cities can drive settlement patterns, shift dangerous land use away from high-risk zones, build strategic transportation corridors, and create safe places for recreation. Opportunities for climate mitigation are plentiful: for example, good zoning plans can promote walkable cities, thereby reducing vehicle traffic and lowering emissions. Similarly, many adaptation objectives can be achieved through zoning. Those same walkable cities contribute to lower local temperatures and improved fitness for residents. Developing new approaches to the process of zoning is also an opportunity to improve decision-making patterns by directly engaging neighborhood leaders and building mutually beneficial relationships with key stakeholders. Planning, then, is a complex and critically important adaptation tool for cities and can offer both process and outcome gains. Indeed, a direct relationship has been found between a city's success with urban planning and its ability to adapt.[6]

Effective land-use planning in cities has to begin with accurate data. Planners need to understand current land-use patterns, variable threats facing different neighborhoods, demographic layouts, and projected new uses, among other things. Most planning efforts approach this challenge with either a hazard-based or a vulnerability-based framework for analysis. Hazard-based planning identifies areas with the highest exposure to hazard and then prioritizes the reduction of risk by addressing exposure. A city might, for example, identify neighborhoods directly exposed to sea-level rise through this methodology; the solution could be the construction of hard barriers or perhaps the development of an integrated combination of new concrete seawalls and NbS (see chapter 3) like wetlands expansion. City planners could go even further, potentially prohibiting any more residential development in those high-risk zones. Vulnerability-based planning, by contrast, identifies neighborhoods that lack coping mechanisms or otherwise are highly susceptible to harm and then prioritizes service delivery to reduce vulnerability. Using this approach, a city might, for example, identify squatter neighborhoods as highly vulnerable; solutions could include improving living conditions by making education and health care more accessible, mandating more resilient building materials for shelter, and relocating residents away from high-risk zones (see the discussion of managed retreat in chapter 7).

It may seem intuitive to consider both hazards and vulnerability in planning, as the two dimensions are intertwined. But the reality is that cities have to prioritize allocation of scarce resources, and each of these approaches to zoning sets the planning team on a different pathway toward addressing the adaptation challenge. In addition, each faces daunting implementation difficulties and may even be unintentionally self-defeating. For example, in Mumbai, urban planners seeking to relocate thousands of vulnerable people living in informal settlements quickly realized that their own new zoning laws—intended to eliminate new high-risk construction along the coastline—had created a land crunch, rendering it nearly impossible to establish new housing for those in need.[7] Many urban areas face related challenges, with population densities that strain against limited space. Without room to expand or redistribute populations, cities are constrained in how much they can accomplish through zoning alone.

Experts caution against interpreting these challenges as a call to make cities less dense. To the contrary, they note, urban density offers many advantages for society, including efficient heating for buildings, professional and educational opportunities, and progressive transportation systems. Instead of reducing density, some cities have gotten creative. Combining zoning with new designs allows urban leaders to reimagine how humans can coexist with the built environment. NbS, for example, have been shown to be particularly effective for addressing heat islands in cities. Through stakeholder identification and community engagement, the NbS decision-making and implementation processes themselves "provide a democratic entry point to addressing many urban challenges,"[8] thereby creating avenues for coupling urban planning with capacity building. Green roofs, another example of NbS, are relatively cheap to install, making them widely accessible. At all levels of society, residents can come together to create opportunities to install them, and once in place, they offer co-benefits for small-scale food production, new spaces for recreation, increased storm-water capture, and creation of habitat for biodiversity. In these ways, dense populations become an asset rather than a liability.

Reform in urban planning can extend beyond zoning to include permitting requirements for new construction. Builders might be required to describe the materials they will use, the process they will follow for securing those materials, and other operational matters; if the city is prioritizing climate-smart development, it may require new buildings to incorporate shading, passive ventilation, thermal massing, photovoltaics, green chemistry, and waste-heat recovery and reuse systems. Passive house design, which originated in climate mitigation efforts, incorporates these attributes as a way to reenvision the built environment as less dependent on the consumption of fossil fuels. Climate adaptation aspirations are also addressed by passive house strategies, as a cooler home can help residents adapt to hotter temperatures without relying on emissions-intensive air-conditioning. Overall, the trend toward adaptive building codes reflects a larger sense that "the era dominated by sealed building envelopes, pervasive impervious hardscapes, standardized industrial materials, and profligate use of fossil-fuel-powered mechanical ventilation can and should end."[9] For these building and design tools to spread more widely, cities must enact legally binding building codes that support the new standards. With only sixty-five countries requiring urban planning department approval before a developer can obtain a construction permit, it appears many governments are missing this opportunity.[10]

Policies at the national level can enable these shifts. Intentional decentralization, for example, can empower urban leaders to take action. Combined with technical and financial support, this kind of nested authority structure can have an energizing impact on municipal governments. Efforts to build that technical capacity come from a range of regional and global institutions—including the African Adaptation Program, C40 Cities, the Covenant of Mayors, and the former Resilient Cities Initiative at the Rockefeller Foundation—that have emerged to connect city leaders with one another. In 2018, the IPCC Cities and Climate Science Conference focused on helping those leaders figure out how to work more closely with their residents. The starting point was a commitment to a process of coproduction of knowledge, where experts work alongside locals and decision makers integrate that information in the design of city landscapes across sectors and scales. Sharing best practices and lessons learned through these venues can contribute to the creation of templates for urban adaptation and, ultimately, to improved outcomes.

ADAPTIVE CAPACITY

Let's dig deeper into the notion of adaptive capacity (see the introduction). One of the most vexing aspects of urban adaptation is the many uncertainties about which hazards present the most urgent threat. Can cities reasonably use urban planning processes to simultaneously plan for floods, droughts, wildfires, and food shortages? Do the same updated building codes that offer respite from heat also succeed with unexpected weather conditions like excessive rainfall? Will a hazard-based urban plan unintentionally compound existing vulnerabilities instead of

CASE STUDY: ANTWERP

When it was discovered that the Belgian city of Antwerp (population one million) was suffering twice as many dangerously hot days as the surrounding countryside, urban planners adopted a suite of new measures.* Drawing on data from advanced climate modeling tools, they began by holding stakeholder meetings to connect scientists with local residents; those discussions sought to align technical thermal assessments with perceptions of heat stress. The resulting building codes now require builders to incorporate a green roof if possible, choose permeable materials for the construction of parking lots, and paint all new structures in light, sunlight-reflective colors. Today, the city is able to forecast five days of weather at the neighborhood scale, allowing it to build a robust EWS that issues heat-wave alerts to vulnerable residents.

BOX FIGURE 4.1 Locator map of Antwerp, Belgium.

City leaders note that up-front costs associated with these layered measures were high, and some elements of the program are still in development; still, long-term maintenance expenses are projected to be low, and once implemented, the adaptation strategies will be in place indefinitely. Antwerp's leaders believe that in addition to reducing tangible risk, the city's adaptation policies have strengthened the collaborative capacity of its administrative departments and raised public awareness of climate change.

*Climate ADAPT, "Adapting to Heat Stress in Antwerp (Belgium) Based on Detailed Thermal Mapping," Case Studies, last modified September 30, 2021, https://climate-adapt.eea.europa.eu/metadata/case-studies/adapting-to-heat -stress-in-antwerp-belgium-based-on-detailed-thermal-mapping.

alleviating them? Even the most robust risk-averse zoning plans and climate-smart building codes cannot provide effective adaptation comprehensively.

A city's direct coping strategies—zoning plans, regulations, strategic plans, and policies—will be amplified when matched with indirect coping mechanisms like durable social cohesion, lack of political corruption, and strong land tenure.[11] Planting green roofs should be but one prong in a more comprehensive strategy to address heat effects. Other elements of that strategy might include improved access to emergency cooling centers in vulnerable neighborhoods, increased investments in health care, and EWSs that are activated through an interagency process infused with attention to the needs of highly vulnerable residents. Building institutional pathways that can accomplish these interconnected goals is essential and requires overcoming entrenched government silos. Thus, for any isolated adaptation measure to succeed, the larger goal must be the creation of flexible, informed, empowered, agile, and motivated governments; these are all attributes of adaptive capacity.

At its core, building adaptive capacity is both a top-down and a bottom-up process, where scientists and policy experts engage locals to determine which risks matter, where those risks exist, and how to reconfigure adaptation strategies to best achieve protection. Investing in these stakeholder engagement efforts contributes to adaptive capacity in two ways: the outcome—a new zoning plan or a revised city strategic plan, for example—is valuable, but the very process of building that plan is itself a form of capacity building. Central governments should not be entirely sidelined during this process, as they have essential responsibilities for implementing enabling legislation and offering both funding and technical support resources. To be clear, building adaptive capacity should be prioritized at all layers of government. The topic is described here because urban centers represent particularly vibrant opportunities for growth through planning processes.

PROGRESS: INDICATORS AND REPORTING

How are cities faring in their efforts to adapt? So far, unevenly (see, for example, figure 4.2). Measuring adaptive capacity is notoriously elusive, and many policies with benefits for urban adaptation are difficult to track. Remember that unlike climate mitigation, which can be tracked with a single consensus metric—GHG emissions—adaptation defies easy quantification. Indicators for adaptation are still being developed, and metrics for urban progress are similarly nascent. For now, case studies and qualitative assessments form the bulk of what's known about cities' progress around the world. One study relied on cities' self-reported implementation of adaptation initiatives

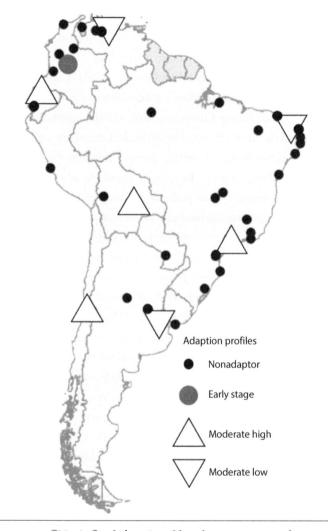

Adaption profiles

● Nonadaptor

● Early stage

△ Moderate high

▽ Moderate low

FIGURE 4.2 Cities in South America—like urban centers everywhere—have made varying degrees of progress toward climate change adaptation planning and implementation.

Source: Image adapted courtesy of Malcolm Araos.

to conclude that only 15 percent of cities with more than one million residents report any such success.[12]

In the EU, the most heavily urbanized region in the world, direct attention has been paid to urban dimensions of adaptation. Many practitioners turn to the Climate ADAPT website, where the Urban Adaptation Map Viewer collects information on climate threats facing European cities and the adaptation actions undertaken in each one. Urban leaders regularly update their efforts through an interactive portal, enriching the data that's made available to peers. Keeping in mind the adage "What gets measured matters," it is perhaps not surprising that adaptation progress in that region has been substantial.

In African cities, the picture is bleaker; a recent study found that despite the existence of a handful of high-profile adaptation projects that boast political support, across the continent "there is little evidence to suggest that substantial adaptive approaches are being effectively implemented."[13] The main culprit for the breakdown between planning and implementation appears to be pervasive weakness in building and measuring adaptive capacity. In the decades since independence, sub-Saharan African countries have been beset with profoundly weak political institutions, overly centralized government structures, and limited access to financial support. The City Resilience Program, launched by the World Bank in 2017, is designed to empower cities to attract investments, but difficult funding challenges remain in many cities. We return to the question of adaptation funding in chapter 8.

The lack of cross-sector comparative indicator data for urban adaptation has important implications. Without reliable monitoring, cities are unable to track their own progress, share results with peer metropolitan districts, or raise funds. Indeed, the lack of monitoring and evaluation is itself a barrier to adaptation. Efforts are underway, both regionally and globally, to remediate this gap.

For example, the C40 Cities network, comprising ninety-six of the world's largest cities, has developed the Climate Change Adaptation Monitoring, Evaluation, and Reporting Framework (CCA MER). Each participating city is urged to track adaptation actions, outputs, outcomes, and impacts, thereby measuring the effect of a single policy across conceptual dimensions (figure 4.3). Each component is matched by quantitative indicators, such as the number of people displaced by flooding before and after a new water storage facility was installed. As cities begin to embed these indicators in their own urban planning documents, municipal leaders will be more accountable to residents. Peer cities can begin to learn from one another, flagging implementation hiccups and sharing strategies for overcoming them. Importantly, the C40 network does not collect these data directly; rather,

	Action	Output	Outcome	Impacts
Rainfall	Convert recreational and open spaces to water squares and parks	Additional water retention area	Reduction of floods due to rainfall	Reduced exposure to flooding
Storm surge / sea-level rise	Installing floodgates	Floodgates installed	Reduced storm surge flooding	Reduced exposure to flooding
Heat wave	Increase shade in public areas	Shading structures installed	Moderated temperatures	Reduced exposure to heat wave
Drought	Rainwater harvesting	Rainwater collecting system installed	Increased water availability	Reduced vulnerability to drought
Wild fires	Implement preventive forestry management	Controlled burns	Reduced wildfire events	Reduced vulnerability to wildfires

FIGURE 4.3 Linking policy actions with short-term outputs, longer-term outcomes, and impacts on stakeholders, the C40 Cities' CCA MER showcases the power of indicators to promote accountability and peer learning.

Source: C40 Cities Climate Leadership Group, "Measuring Progress in Urban Climate Change Adaptation," 2019.

it works to build capacity within its member cities so they can develop their own indicator systems with locally appropriate data collection strategies.

CONCLUSION

As centers for education, finance, and technology innovation, cities are well positioned to build adaptive capacity, and they can serve as shining manifestations of how nationally scaled adaptation policies can connect local communities with scientific experts. Effective adaptation initiatives in urban areas contribute a range of benefits for cities that extend beyond the intended immediate objectives, and a safer, more versatile, better-engaged populace makes for a better city, with or without climate change.

Next, we turn our attention to the many ways these vibrant urban settings are dependent on healthy rural landscapes for sustenance. Agriculture and land-use conversion may take place out of sight for many urban residents, but without reliable food supplies, even the most progressive and adaptable urban communities will struggle. Those dynamics form the next layer of climate adaptation governance to be explored here.

5

AGRICULTURE, LAND USE, AND FOOD SECURITY

RURAL LIVELIHOODS

Amid trends toward urbanization, rural areas remain essential building blocks of modern societies. Rural communities are characterized by lower-density settlement patterns and distance from urban cores. On average, rural dwellers are older, less educated, and less healthy than their urban counterparts; they are also more likely to rely on the informal economy than those who live in cities.[1] Rural landscapes are the centers for agriculture, fishing, and food production and are home to the most productive habitat on the planet, making them hot spots for biodiversity. In some countries, rural areas are also hubs for tourism and recreation. Without urban institutions to structure social organization, many rural residents rely on extended kin networks for social cohesion.

Most indigenous communities are situated in rural areas, mirroring generations of resource-based livelihoods. Rural inhabitants tend to be politically marginalized, as they reside far from centers of power, but this is not true everywhere: it's worth noting that in some countries, like the United States and EU member states, the agriculture lobby is powerful, granting high visibility to a relatively small rural subset of the national population.

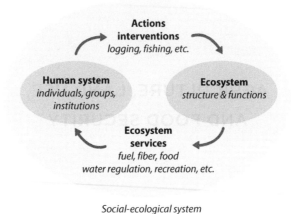

Social-ecological system

FIGURE 5.1 Human systems and ecosystems are inextricable from one another, creating both barriers to and opportunities for improved climate adaptation.

Source: Adapted from Resilience Alliance, *Assessing and Managing Resilience in Social-Ecological Systems: Supplementary Notes to the Practitioners Workbook, Vol. 2* (2007). Image adapted courtesy of Esteban Jobbágy.

The social and natural system attributes described earlier combine to form a self-reinforcing socioecological system (figure 5.1). Indeed, so intertwined are the fates of our ecological systems and our human societies that protecting natural processes on land can have important consequences for social adaptation. Researchers explain it this way: "Rural areas are 'biocultural refugia' that safeguard not only biodiversity toward global food security but also the culture-specific practices that maintain that biodiversity."[2] Here we have treated rural areas as a distinct form of human settlement; in reality, of course, urban and rural patterns exist along a continuum, with suburban and exurban areas complicating this dual structure.

AGRICULTURE AND FOOD SECURITY

Today, there are an estimated 570 million farms on the planet, with the vast majority—90 percent—being small and run by an individual or family.[3] Despite global patterns of increased urbanization (described in chapter 4), perhaps invoking images of hollowed-out rural areas, the world's farms have succeeded in increasing food production over time. Land-use change—plowing through natural ecosystems to make way for cultivated agriculture—is one driver of this pattern, as more farmland can produce more food. Improved seeds, inputs, and technology—a suite of reforms known broadly as the Green Revolution—make up another. Since 1961, these trends have caused the total production of cereal crops to increase by 240 percent.[4] Food demand has risen in step with this increased yield, but pervasive pockets of hunger suggest that the sheer quantity of food is not the main problem. Today, an estimated 690 million people are chronically undernourished, and another 2 billion are overweight or obese;[5] both are forms of malnourishment. The reliable distribution of nutritious and affordable food—a concept called *food security* (figure 5.2)—remains a pressing global challenge that affects people in both urban and rural living environments.

Climate change threatens all elements of food security and can have devastating effects on efforts to alleviate hunger. Some of those impacts are direct, as agriculture is especially susceptible to erratic rainfall. Not only does fluctuating precipitation lead to reduced yield, but also it drives deep uncertainty among farmers, who respond with a range of uncoordinated adaptive responses. Climate variability within growing seasons has already had a substantial negative effect on agricultural outputs in critical farming regions.[6] Researchers predict

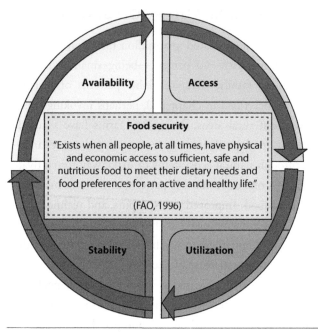

FIGURE 5.2 The four prongs of food security underscore the complexity of today's global food system, an important target for climate change adaptation.

Source: Image adapted from Sandesh Adhikari, "Food Security: Pillars, Determinants and Factors Affecting It," Public Health Notes, 2018.

cereal production will decline by 40 percent by 2050—reversing decades of increase—unless profound changes to emissions alter the business-as-usual trajectory.[7] Even one season of reduced yield can set in motion a cascading pattern: food prices will fluctuate, families may become food insecure when they can't afford groceries, and farmers will face unreliable income flows.

Indirect climate-driven changes can also have profound effects on food security. If a major transportation corridor is flooded, for example, supply chains will be disrupted, farmers will be unable to get their crops to market, and food waste will spike. Many versions of these scenarios exist and are already manifest. It's worth noting that not all the news is negative. Higher-latitude regions have shown increases in yield as previously frigid landscapes warm and new cropping techniques emerge. But even these seemingly positive outcomes may require an adaptive response; surplus produce must be processed, stored, and then brought to the marketplace. In many parts of the world, grain storage infrastructure is weak, and higher yields lead to more food waste. For some countries, trading routes are entrenched and create path dependency, making it difficult to transport new food supplies efficiently to those experiencing hunger.

Impacts from declining food security are not distributed equally. Multiyear droughts have a disproportionate impact on yield in poorer countries, where technological fixes are few, threatening the survival of subsistence farmers. But industrialized, wealthier countries are also at risk. Agricultural policies in the United States, for example, tend to maximize short-term yield through monocropping; this government priority is undergirded by federal subsidies that drive farmers' decisions. When drought hits, those American farmers have not diversified their planting strategies and stand to lose everything.

Autonomous adaptation is already a feature of rural life.[8] Indeed, research shows that those who rely on their crops for survival have significantly greater awareness of climate impacts on agriculture and food supply than nonfarmers.[9] Everywhere farmers voluntarily adjust planting strategies—the strain of crop, the timing for planting, and the use of inputs to support yield—in response to perceived shifts in seasons and weather. Rural workers in agrarian societies in the tropics "are already adapting to chronically hotter temperatures in common ways, such as adjusting when and how they work."[10] In some locations, where farmers have historically planted their crops on a rigid seasonal schedule, researchers have observed shifts to a more flexible planting schedule that allows farmers to manage uncertainty by waiting for the onset of rainfall before planting. Some farmers have built small-scale collection reservoirs to harvest rainwater, deploying an indigenous technique for short-term water storage.[11] Some have increased pesticide use in response to reduced yields.[12] Research shows that while these autonomous adaptation behaviors can lead to significantly improved outcomes, those benefits are highest in areas where farmers have access to reliable information about current and future climate variables, also known as climate services (see chapter 2).[13] But even then, these individual behavior changes are uncoordinated and do not comprise long-term or cross-sector adaptation. They may address short-term hazards but fail to account for longer time horizons. And they occur only in response to anecdotal experience, not projected trends or distant supply chain disruptions.

Systemic, planned adaptation builds from autonomous efforts by mainstreaming best practices and supporting broad deployment through institutions. In agriculture, planned adaptation includes adjusting incentives through national subsidy reform, building ecologically sound irrigation infrastructure for broad deployment, constructing reservoirs for improved water storage, and investing in climate services. All of these approaches require significant government capacity, as farmers live in remote locations, making it harder to reach them and more expensive to provide them with services.

Water supply is a central challenge. Irrigation—the intentional use of water for growing crops—is the most important input for many farmers. It offers a buffer from short-term drought and allows farmers to have some control over how much water their crops get and when. Across much of Asia and the Middle East, approximately half of all agricultural land is irrigated. India irrigates 35 percent of its agricultural land, the United States just 14 percent, and sub-Saharan Africa only 6 percent.[14] Rain-fed agriculture is so vulnerable to climate change that many consider the expansion of irrigation an essential step toward stabilizing food security. With an estimated 50 percent higher yield, irrigated farms are also linked to higher incomes for farmers. Not surprisingly, farmers prefer irrigation above all adaptation policy interventions.[15]

But irrigation comes with some important downsides. Freshwater supply is finite, and population growth creates steady demand for increased food production; already 70 percent of global freshwater is used for agriculture, and diverting additional water to that sector has opportunity costs for residential and commercial use.[16] Irrigation also uses a lot of energy to pressurize sprinklers, run drip systems, and pump groundwater. Burning fossil fuels and tapping into a scarce water supply both contribute to the maladaptive nature of irrigation as an adaptation strategy. Some research even suggests that farmers with irrigation infrastructure are so buffered from the daily reality of rainfall variation that they are less likely to perceive climate change impacts and therefore less likely to undertake coordinated, and potentially costly, adaptation measures.[17] Despite the short-term benefits of the strategy, some international development organizations don't support irrigation policy everywhere.

Water supply is one piece of the puzzle, but comprehensively rethinking all elements of the food system is necessary for long-term adaptation. *Climate-smart agriculture* (CSA) envisions a way to increase productivity while also supporting equitable gains in income, improved food security, and enhanced resilience to climate risk at all levels of society (figure 5.3). The concept is broad and has been applied to adaptation in livestock,

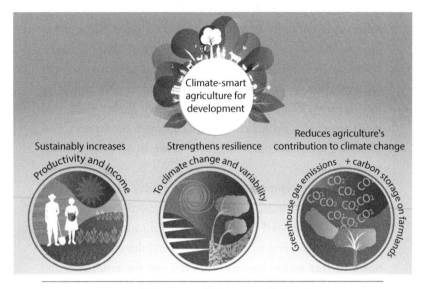

FIGURE 5.3 CSA supports both sustainable development and climate change adaptation goals.

Source: Food and Agricultural Organization of the United Nations. Reproduced with permission.

fisheries, and crops.[18] Promoting CSA more widely will require financial investments, cooperative action, and political will. Some progress is now apparent, as key principles of CSA are being formalized in NAPs and other national-level adaptation planning documents.

LAND USE, PROTECTED AREAS, AND LAND TENURE

Changes in land use that have accompanied human settlement, industrial expansion, and agricultural production have important connections to climate change. Most prominently, the loss of tree cover is considered one of the most powerful drivers of global warming; climate mitigation policies that reduce deforestation are central to balancing a global carbon budget. But there are implications for climate adaptation too. Heat waves and heavy precipitation events can be intensified by locally degraded landscapes, and those same events then inflict new damage. Effective adaptation can intervene in the damaging cycle. Afforestation is perhaps the best-known version of this co-benefit opportunity: increases in tree cover contribute to cooler local temperatures as well as improved air and water quality, suggesting that land restoration—converting agricultural land back to forestland— might be advisable where possible. Still, in the context of threats to food security, taking lands out of production may create difficult trade-offs for decision makers.

Policies that allow land to be protected—as when a government designates an area as a national park, limiting the kinds of uses permitted there—can contribute to climate mitigation efforts by keeping trees standing; blocking new, unwanted land-use change; and supporting healthy ecosystems that serve as carbon sinks. These same areas can support climate adaptation by fortifying ecosystem services like flood control and providing scientists with valuable natural landscapes in which to study ecological system function. The International Union for the Conservation of Nature (IUCN)—an intergovernmental organization (IGO) that helps countries build biodiversity protection strategies, among other things—defines a protected area as "a clearly defined geographical space, recognized, dedicated and managed, through legal or other effective means, to achieve the long-term conservation of nature with associated ecosystem services and cultural values."[19] To date, countries have placed nearly 15 percent of land under some level of protection, approaching the targets set by the Convention on Biodiversity in 2011.[20] National disparities are broad and exhibit regional patterns: Latin America, western Europe, Oceania, and Africa have overall the highest proportion of their land protected (figure 5.4).

But establishing protected areas is not always a purely beneficial solution. Sometimes the imposition of restrictions by governments can undermine local adaptive capacity by reducing access to resources. Indigenous groups have been particularly vulnerable to forcible removal from their land as part of government efforts to create the impression of "unspoiled nature" for visitors.

Share of land area that is protected, 2018
Terrestrial protected areas are totally or partially protected areas of at least 1,000 hectares that are designated by national authorities as scientific reserves with limited public access, national parks, natural monuments, nature reserves or wildlife sanctuaries, protected landscapes, and areas managed mainly for sustainable use.

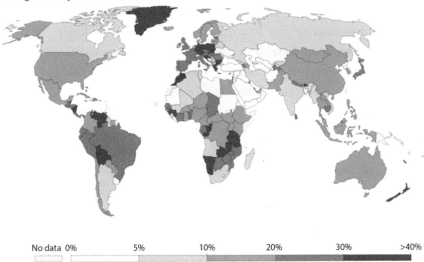

FIGURE 5.4 Establishing protected areas is a national strategy that can contribute to achievement of both climate mitigation and adaptation goals.

Source: UN Environment Programme (via World Bank).

Indeed, even when resettlement isn't explicitly part of the establishment of a new protected area, limiting human access as a way to support ecosystem function has implications for equity and long-term rural viability. Policies that seek to maintain tree cover in a protected forest reserve, for example, may also prohibit agricultural practices that sustain nearby rural communities.

Part of the reason indigenous groups have been susceptible to forcible removal is their relatively weak land tenure. The concept of *land tenure* is comprised of three layers of rights: use, control, and transfer. Strong land tenure means a household has legal or customary title to a parcel of property, has control over what kind of use they choose to pursue there, and has the legal authority to buy, sell, or bequeath that property. But while rural families in many parts of the world may have lived on a parcel of land for generations, they likely lack a legal piece of paper that authorizes them to be there. They are constrained in how they can use the land, since the government maintains control over their property. And they are unable to transfer the land, meaning they have little incentive to invest in long-term sustainability practices on the property and cannot build wealth through the improvement of their home. In this way, many rural communities with deep generational attachment to the land find themselves disconnected from legal systems of land rights, especially when those systems were established through colonial occupation.

Across much of Africa and Asia, European colonists, accustomed to a codified system based on private land rights, encountered unfamiliar indigenous communities with customary systems based on communal land use. So foreign were these systems of organization that the colonists did not recognize them— or they chose to ignore them. Lands occupied by nomadic and hunter-gatherer groups were labeled vacant. Rural subsistence strategies that relied on communally accessible resources were not well understood, and the creation of protected areas served to criminalize essential aspects of these traditional rural lifestyles. These trends have contributed to diminishing rural growth and have pushed people—mostly young, mostly male—to leave their homes in search of urban opportunity.[21] Even for those who remain, weak land tenure is now well known to create or exacerbate conflict. It is linked to high poverty and an inability to accumulate wealth, creating a poverty trap.[22]

In these ways, protected areas, intended to preserve nature and guard humanity against the ravages of climate change, can in fact drive impoverishment and reduced adaptive capacity.

CASE STUDY: RWANDA

The Batwa people comprise less than 1 percent of the total population of Rwanda, a tiny country in East-Central Africa.* Historically, this tribe of hunter-gatherers relied on access to communally shared natural resources for subsistence. But when the government established national parks—seeking environmental protection, habitat for imperiled wildlife, and revenue from tourism—the Batwa were forcibly removed from their traditional sites. They were not compensated or resettled and have been described as "victims of conservation."† During the 1994 genocide, an estimated 30 percent of the Batwa people—mostly men— were killed; the remaining tribe members were mostly women and children. Today, the Batwa are intensely marginalized. They are known throughout the country as potters, and they sell their ceramics at roadside stands, a practice that generates very little income. They have high mortality rates, little access to health care, and almost no formal education, and they remain unrecognized

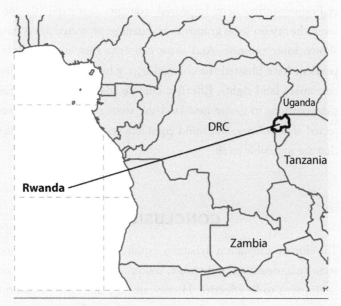

BOX FIGURE 5.1 Locator map of Rwanda.

by their government. Weak land tenure is a defining feature of their circumstances, and in Rwanda's densely populated countryside, they have been largely ignored by government farm policies that have bolstered rural life. In a country that already ranks as one of the most highly vulnerable to climate change, the Batwa are among the most highly vulnerable people on the planet.

*Benon Mugarura and Anicet Ndemeye, "Batwa Land Rights in Rwanda" (micro study, Minority Rights Group International, London, 2002), https:// minorityrights.org/wp-content/uploads/old-site-downloads/download-93 -Batwa-Land-Rights-in-Rwanda.pdf.

†Gerald Tenywa, "Climate Change a Threat to Forgotten People of Uganda," *New Vision*, April 16, 2018).

But solutions exist, and, perhaps, there are legitimately benefi-cial opportunities here. Improved strategies that explicitly con-fer authority on local groups to co-manage protected areas have shown some promise. And some countries that are now inde-pendent have blunted the colonial legacy by legally recognizing customary land rights. Effective climate adaptation will require many nations to devise new land-use strategies, reconsider pro-tected area policies, and build equitable systems for land tenure that are accessible to all.

CONCLUSION

The climate adaptation dynamics explored in this chapter—food security, agriculture, land tenure, and protected areas—all rely on institutions to be effective. Historically, the most robust institu-tions exist in settings with strong organizational, administrative, and fiscal capacity: urban centers and international venues. The national and regional policies developed there serve an essential enabling function for local governments in rural areas, reflect-ing a hierarchical and nested policy structure.[23] But real progress necessitates institutional reform so that rural communities have direct access to power and don't need to rely only on distant sup-port. How can this kind of reform unfold?

Due to the sheer dispersion of homes, a participatory strategy for rural adaptation is essential. Consultative work-shops conducted by decision makers should be bottom-up; the process can create awareness and build policies that benefit from local support. One researcher, studying a community-led adaptation process in Germany, describes the pattern this way: "It is not the representativeness but rather the motivation and engagement of farmers that [is] decisive for successful changes in behavior."[24] Building participatory processes is one concrete way local leaders—backed by national enabling policy and funding support—can invest in human-capacity development. Farmers' associations, agriculture extension offices, and community centers are all examples of localized institutions that can be activated through a process of focused stakeholder development. This kind of institutional network will also increase the likelihood of productive communication between national and local leaders that is aligned with the decentralization principles discussed in chapter 1. Adaptive capacity is firmly rooted in this kind of equitable institutional framework.

While these layers of institutional reform take root and rural communities build the capacity to drive their own adaptive futures, shorter-term risks to agricultural livelihoods persist. Floods, drought, and a host of other departures from historical weather variation threaten the viability of farms around the world. Agricultural producers are increasingly embracing insurance as a tool to buffer them from these shocks. The next chapter explores how some new insurance products are intended to support climate adaptation and build resilience.

6

INSURANCE AS RISK TRANSFER

OVERVIEW

If adaptation to climate change is essentially an act of assessing and managing risks, then finding ways to buffer shocks from extreme weather is of primary importance. Insurance has long functioned in exactly this way, building pools of individuals—or, for underwriters, pools of companies or countries—to spread out the risk across society. If one person falls ill, is injured, is the victim of a robbery, or suffers other harm, the shared pool of resources is deployed to the sufferer. Historically, insurance was available only following a natural disaster or theft. Now, a range of insurance products is seen as essential for building resilience ahead of predictable climate-change-fueled disasters.

The concept of risk pooling is timeless and rudimentary. Indeed, rural communities have long practiced versions of this strategy; for example, across farming communities in Africa, households build up savings in good years in part to enable them to support fellow farmers in lean years.[1] These traditional forms of loss spreading highlight the ways the mere existence of insurance—separate from any claim—can drive human behavior through incentives. For example, when a farmer knows his community is prepared to support him if his crops fail, he may be more likely to invest in newer and perhaps higher-yield approaches to planting, even when they are unfamiliar and therefore risky. Without the risk transfer mechanism in place, more farmers would be unwilling to experiment. In this example, we might consider insurance to have a positive impact on farmer behavior. But what if that farmer lives on marginal lands or in an area with chronic flooding? Without insurance, he might decide to relocate his family to more productive farming land, a move that would be consistent with long-term adaptation. But with insurance, he might choose to stay on his low-producing plot and survive through risk pooling. In this example, insurance functions to obscure the limitations of current livelihood practices; it even runs into *moral hazard*—here understood as a form of maladaptation—in which the costs of the farmer's unsustainable choices are not his to pay.

Insurance also functions as a price signal for developers. If private insurance companies pull out of a region—as they have done in some places on the frequently flooded East Coast of the United States—it's an indication of asset exposure. For potential real estate developers, this likely means building there is a bad investment. At the aggregate level, variations in insurance pricing and availability can powerfully drive settlement patterns and private investment choices—and potentially lead to government regulations that address those emerging risks. But experts note that equity concerns can interfere with these price signals. For example, if insurance rates increase, perhaps accurately reflecting flooding risk, only wealthier homeowners will be able to afford those policies; one result would be a gentrified neighborhood with high-income residents. That new neighborhood wouldn't function as a clear sign of impending risk. To the contrary, those high-value homes would likely be misinterpreted to signify safety.

In these ways, insurance is a complicated tool for building resilience. With only 3 percent of total asset losses in low- and low-middle-income countries insured, expanding insurance availability is an important goal for many governments that are unable to cover the cost of increasing damage associated with natural disasters. In 2018, the InsuResilience Global Partnership reported that 33.2 million poor and vulnerable people were protected by insurance globally; by 2019, the number had more than doubled, to 89.4 million.[2] But more insurance is not necessarily always better. This chapter explores how insurance can best be deployed to advance climate adaptation.

RISK POOLING: SCALE AND KEY CONCEPTS

Two kinds of insurance, indemnity and parametric, offer different ways to pool risk. Indemnity insurance is familiar to many who seek to protect a piece of valuable private property. An asset with a known value—an heirloom necklace, for example—can be insured, with the insurance premiums tied to the likelihood of that asset's loss and the cost of replacing or repairing it, should it be damaged. If the necklace is stolen, the policyholder files a claim, and an agent from the insurance company inspects the damage in person or seeks police verification of the loss. Two features of indemnity insurance make it a poor fit for climate adaptation: it responds only to losses that have already happened (rather than anticipating those losses), and it covers only the insured asset. Indemnity insurance is common across wealthier countries, where more people own valuable assets and have the cash flow to protect them.

Parametric insurance—also called index insurance—works differently. Not tied to any particular asset, parametric policies respond to catastrophic events. When a predetermined data point is reached—centimeters of rainfall within one week across a defined geographic area, for example—the product is automatically triggered to pay its policyholders for damage. No insurance agent has to visit the property, thereby lowering transaction costs. Index-based policies generally offer lump-sum payouts to farmers in areas hit by drought or flood, and these farmers are free to use their payouts any way they choose. Policies providing forecast-based payouts even allow the release of funds ahead of the predicted weather event, giving farmers a running start. This kind of insurance is much better suited to support climate adaptation and is much more common in developing countries with low insurance penetration. With more stable cash flow, farmers can leverage their insurance coverage to help access credit and accumulate wealth. Indeed, "for many, an insurance contract can be a more dignified and secure means of coping with disasters than dependency on the ad hoc generosity of donors."[3]

Scale matters. Parametric insurance products are available to individuals and small businesses, but the companies offering these micropolicies have trouble reaching enough individuals to make them affordable, and gaps in coverage are pervasive. Most low-income countries have insurance penetration under 1 percent.[4] At a larger mesolevel, groups of individuals—usually farmers, gathered in a collective or an association—can find products tailored for their groups. In those settings, individuals are still the beneficiaries of the payouts, but the policies are held at the group level and administered through the collective. At a macroscale, country governments can obtain insurance coverage to help them with large-scale disaster-related expenses, including payouts to affected residents. Some states have also joined together to pool risk regionally. For example, the Caribbean Catastrophe Risk Insurance Facility was established to provide immediate payouts in the face of disaster. Today, twenty-two countries participate.[5] The African Risk Capacity is another example. In 2019, the agency's parametric model enabled it to alert the Senegalese government that rainfall was scarce and forecasts suggested continued drought. As agreed, this metric triggered a payout of US$23 million to the country several weeks before the end of the farming season, allowing farmers to take steps to protect their livelihoods.[6]

Even the most robust, layered system of insurance coverage runs into liquidity problems when a major disaster strikes. Reinsurance, a process in which insurance companies themselves are insured, may be necessary to cover claims in the event of multiple simultaneous losses. Reinsurers are usually funded through the private marketplace but can also be supported by governments (figure 6.1). Most reinsurers use risk diversification as a tool for spreading exposure, meaning the companies are potentially offering a range of policies to different clients to cover fire, flooding, and hurricanes across different geographies. These policies are essential tools that not only protect the companies' bottom line but also help them demonstrate solvency for government regulators.

For any of the parametric products described here to function, having reliable weather data is a necessary precondition. Insurance companies must have a way to assess past weather patterns in order to determine trigger points, and then they need reliable real-time information to assess when those trigger points have been reached. Customers also need access to weather data, both for planning their own actions and for holding insurance companies accountable. The provision of access to reliable data sets covering past, present, and future dimensions of climate is known as *climate services* (see chapter 2). Forecast-based financing (FbF; figure 6.2) relies heavily on climate services, which are used to establish triggers for early payouts; with sufficient advance

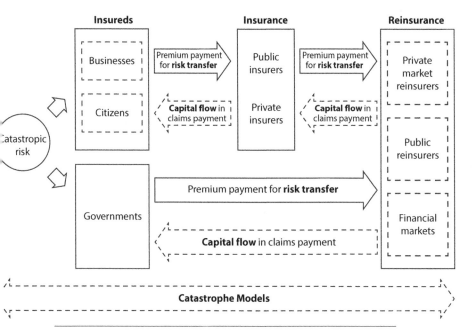

FIGURE 6.1 Insurance products spread risk across both the public and the private sectors, creating essential buffers for climate risk.

Source: Adapted from P. Jarzabkowski, K. Chalkias, D. Clarke, E. Iyahen, D. Stadtmueller, and A. Zwick, "Insurance for Climate Adaptation: Opportunites and Limitations" (Rotterdam: Global Commission for Adaptation, 2019).

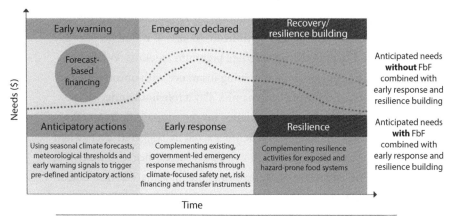

FIGURE 6.2 FbF releases funds ahead of an emergency, empowering policyholders to take anticipatory action and enhancing recipients' capacity to absorb a shock.

Source: © 2019 World Food Programme.

106 Insurance as Risk Transfer

warning, farmers can anticipate an upcoming drought, for example, and use their payouts to buy food for their cattle. Without such coverage, the same farmers will be forced to absorb the losses from the weather event and seek recompense afterward, when it is likely too late to build adaptive capacity.

PRICING AND BASIS RISK

All insurance policies begin with an assessment of risk—defined as a function of hazard, exposure, and vulnerability (see the introduction)—to establish coverage and pricing. If that assessment is accurate, pricing should be a reasonable proxy for risk. The practical application of this principle, called *risk-reflective pricing*, requires people who live in riskier areas to pay more for insurance coverage. But adverse selection—the tendency for people at particularly high risk to purchase insurance at higher rates than those with less risk—can distort the pool and lead to high premium prices. Further complicating the factors that contribute to pricing, decisions about whether or not to enroll in an insurance program don't appear to conform to rational-choice economic modeling.[7] This makes it difficult to use a willingness-to-pay method to establish appropriate premium amounts. When farmers make choices about obtaining insurance with uncertain future benefits, complex social dynamics play a role in their decisions.

Basis risk occurs when claims do not align with losses. Positive basis risk happens when a payout is triggered even when the disaster has not led to correlated losses; negative basis risk happens when losses are extreme, but payouts are insufficient. Indemnity insurance is more susceptible to basis risk, but parametric insurance can also run into payout problems, since the same index formula applies to all farmers in a given geographic location regardless of specific losses.[8] Reducing basis risk is difficult when rates are determined exclusively through climate modeling that is rooted in historical weather trends; when those past weather patterns no longer serve as useful guidance for current and predicted climate, basis risk increases. Technology can help. Remote sensing, for example, has enhanced climate services and improved predictive capacity. The African Risk Capacity has found that integration of these data can help reduce basis risk.[9]

ROLE FOR GOVERNMENTS

In developing countries with low insurance penetration, governments regularly contribute financial, technical, and regulatory support to improve access to insurance coverage. But the involvement of public actors in the provision of insurance raises difficult questions. For example, when a national government steps in to subsidize agricultural insurance for rural farmers' collectives, most agree that the public-private partnership is both appropriate and effective. Similarly, when a government interferes with pricing schemes to even out volatility as a tool for creating larger risk pools, some support those efforts. But when a government underwrites policies that enable residents to persist in risky locations, that changes the equation. In those cases, the government might be contributing to perverse incentive structures, misleading price signals, and negative progress on adaptation.

Two examples from the United States highlight ways in which government subsidies can create maladaptive outcomes. The Federal Crop Insurance Program (FCIP) grew from a largely privately funded program intended to reduce federal payouts following disasters; the program has now expanded to occupy a central place in agricultural policy in the United States. Farmers have remained in high-risk areas because they have insurance coverage, and with only a narrow margin of economic security, they are locked in a system that values commodity crops over other systems of planting that might be more responsive to emerging climate threats. Insurance premiums are established through analysis of past risk, and government subsidies have shielded farmers from experiencing real-time variability. Researchers conclude that "the government effectively spends significant amounts of taxpayer money to subsidize the riskiest of crops, blunting farmers' incentives to choose less-risky crops."[10] With disproportionately strong political power, the agriculture lobby in the United States has been notoriously resistant to reform. The cost of maintaining the FCIP has now reached $12 billion annually.[11]

Another example from the United States—the National Flood Insurance Program (NFIP)—highlights similar patterns. The NFIP was created in 1968 to address decades of flood damage along the vast U.S. coastline, and soon participation in the program was made mandatory for coastal residents. By 2009, more than five million policies had been issued to people living in Special Flood Hazard Areas. The program was untenable even then, but increasing coastal flooding associated with climate change has rendered it nearly bankrupt. Some observers argue that the continued availability of policies in flood zones has "promoted a counter-adaptive psychological world view,"[12] as property owners now see flooding as normal and expected and not as a reason to relocate. Tested by big storm events like Hurricane Katrina in the Gulf of Mexico (2005) and Superstorm Sandy in the Northeast (2012), the NFIP undertook significant reform efforts, including purchasing reinsurance and launching catastrophe bonds (described shortly); in 2019, it announced plans to shift to risk-based pricing.

These high-profile programs in a wealthy country carry conflicting lessons. Government-subsidized insurance can provide institutional stability, promote equitable access to risk transfer, and contribute to public confidence, but it too often leads to maladaptation. Some argue that if the government got out of the insurance business, the private market would potentially reflect actual risk; in this kind of restructuring, the role for governments would be restricted to providing an enabling regulatory framework for markets. Still, without subsidized flood insurance, lower-income coastal residents would face difficult financial circumstances. In this free-market version of the insurance industry, only rich people could afford seaside living, and widespread gentrification of coastal communities would ensue. This tension between equitable access to social safety nets and risk-reflective pricing is evolving in today's context of a rapidly changing climate.

Complex financial instruments such as catastrophe bonds and resilience bonds—which pay out for insurance companies when a natural disaster strikes—have emerged in this context. They may help to address the tension just described by fortifying insurance companies through a government-backed pathway for private investors. Resilience bonds can also shift incentives, as they may lower premiums when climate-resilient interventions are applied to an asset of value. The first resilience bond was issued in Europe in 2019, and observers expect the demand for such instruments to grow rapidly.[13]. For additional coverage of adaptation finance, see chapter 8.

INSURANCE AS ADAPTATION:
KEY CHALLENGES AND BEST PRACTICES

As a strategy to pool risk and build a buffer for vulnerable communities facing climate shocks, insurance is likely to play a vital role in the development of a more resilient society. But the tool also can lead to a range of pitfalls, as we have described. If understood only as a way to rebuild following a disaster, insurance cannot meaningfully contribute to adaptive capacity. Perverse incentives, adverse selection, and the risk of moral hazard bedevil providers. As insurance provision has increasingly made its way into mainstream adaptation planning, lessons have been learned that can help minimize these pitfalls.

One area of interest is asymmetrical gender dynamics. Women—more likely to be homebound with child care and household responsibilities and less likely to be educated—have a higher likelihood of dying in a disaster, although these findings have varied considerably by location. It is perhaps not surprising that women also have different preferences for insurance products, likely caused by different levels of trust in institutions and by relative differences in financial literacy; for example, researchers in Bangladesh report "significant insurance aversion among female farmers."[14] These results suggest that simply offering insurance is insufficient; it must be preceded or accompanied by training in financial literacy. If expanding the risk pool is socially desired, then a more nuanced understanding of potential customers is essential.

The gender dynamic also highlights ways in which insurance products must be developed with the end user in mind. Practitioners note that when private companies decide how much insurance people need, they tend to come up with products that don't fit. Instead, a more participatory process that includes insurance providers, scientists with expertise in climate services, and local communities can help with both uptake rates and the design of appropriate products. Global institutions like the InsuResilience Global Partnership, launched in 2017, have focused on insurance needs in developing countries and urge this kind of bottom-up planning.[15] Building financial literacy is essential, and companies have become creative, integrating simulation games into community meetings to improve understanding and foster adaptive capacity.[16] The lessons learned in those contexts are valuable for all countries. Some best practices are summarized here.

First, insurance best contributes to adaptation when it is integrated with other DRR efforts. That means the risk models used to formulate pricing structures in the insurance industry should be aligned with the models used by development banks and governments and should be connected to climate services and EWSs. Local residents should have access to both information and credit so they can understand and purchase appropriate insurance. If they are developed correctly, insurance products can help build technical capacity, as countries that form partnerships with insurance companies can better understand and manage their risk.[17]

Second, the insurance products themselves should be structured to reduce moral hazard. Generally, experts agree that parametric insurance is less susceptible, since claims are independent of losses, and some go as far as to suggest that "moral hazard is eliminated altogether with index-based or parametric contracts."[18]

CASE STUDY: BANGLADESH

Situated almost entirely below sea level, Bangladesh has been something of a poster child for climate change adaptation for decades.* Sea-level rise is the primary driver of displacement for the 25 percent of the country's population that resides on the coast. Recurrent flooding in the Kurigram District has been particularly difficult, contributing to unstable livelihoods and an ongoing human toll. In 2015, the World Food Program (WFP)

BOX FIGURE 6.1 Locator map of Bangladesh.

partnered with Bangladesh's Ministries of Disaster Management, Finance, Meteorology, and Flood Forecasting to build a stronger FbF system. EWSs are key for the functionality of such a system, and the partners had to agree on appropriate triggers for action, communication strategies for reaching remote villages, and early actions that could reduce risk. Including local stakeholders in the planning process has already contributed to widespread buy-in for the system, and, indeed, vulnerable farmers stand to benefit the most from its implementation. Villagers receive both weather alerts and cash transfers through their mobile phones based on forecasts that are measured against preestablished triggers for river flooding. Integrating FbF tools with existing social protection systems and disaster management planning has enabled the coalition to build a comprehensive risk management strategy. The Bangladesh example supports efforts to scale up such programs through the United Nations Risk-Informed Early Action Partnership (REAP).

*World Food Program (WFP), *Forecast-Based Financing (FbF): Anticipatory Actions for Food Security* (Rome: WFP, 2019), https://docs.wfp.org/api /documents/WFP-0000104963/download/; Damien Joud, Thorsten Klose -Zuber, Alexandra Rüth, and Manuela Reinfeld, "Reducing Disaster Risk Vulnerability in Bangladesh—Partner Perspectives: Summary Points, Questions and Answers" (from a webinar, Food and Agricultural Organization of the United Nations, Rome, December 6, 2018), http://www.fao.org/3/I9733EN /i9733en.pdf; Red Cross Red Crescent Climate Centre, "Risk-Informed Early Action Partnership—'REAP'—Launched at #climateaction Summit: 'Let Us Work Together for a Safe World for Our Future Generation,'" ReliefWeb, United Nations Office for the Coordination of Humanitarian Affairs, September 23, 2019, https://reliefweb.int/report/world/risk-informed -early-action-partnership-reap-launched-climateaction-summit-let-us-work.

But even more-traditional flood insurance can be improved by requiring, for example, resilient design in rebuilding efforts. A policy can also specify ways for customers to reduce premiums through focused investments in risk reduction, thereby incentivizing behavior changes over the longer term.[19]

Third, governments should tread lightly in the insurance business. As described in this chapter, a competitive insurance market operating in the private sector without government intervention may be best positioned to succeed across a variety of relevant criteria.[20] But in many developing countries, a short-term public subsidy is essential, as it can help newly developed insurance products gain a foothold, build community trust through affordability, and contribute to stability in pricing. Evidence suggests that if those subsidies outlive the short time frame, they become increasingly difficult to untangle; the U.S. examples presented earlier are cautionary. The best role for governments over the long term appears to be regulatory, with the development of enabling institutions like mobile banking and secure supply chains.

Last, we must confront the Samaritan's dilemma: after a disaster, most governments will help everyone who hasn't taken preventative measures, especially the uninsured, raising this key question for residents: "Why should they pay the premium for private insurance or invest in self-insurance or self-protection measures if they enjoy a similar amount of protection from the government free of charge?"[21] This is, by all accounts, a difficult problem to solve. Mandatory insurance is one solution, but such a policy immediately runs into an expanded role for government. Perhaps there is a narrow pathway through this challenge, with incentives structured to motivate residents to pursue adaptive behaviors and with companies positioned to reward them for those actions through more affordable insurance policies. New financial tools like resilience bonds can help, and updated programs like the United States' Community Rating System are intended to guide municipalities toward actions they can take to reduce flood insurance premiums for residents. Still, this dilemma underscores one of the most perplexing factors in the effort to better integrate insurance as a tool for climate adaptation: temporal mismatch. All insurance products are short-term by nature; climate change adaptation is a long-term process. Better design, implementation, and cohesion in the insurance industry can maximize its effectiveness.

CONCLUSION

Even under the ideal conditions described here, where insurance is affordable, available, and effective, climate change means some previously viable settings are simply becoming uninhabitable. Increasingly, individuals and communities have had to consider relocating in search of better living conditions. They may move individually, or they may find their entire community targeted for government-led managed retreat. The next chapter addresses this pattern.

7

MIGRATION AND MANAGED RETREAT

OVERVIEW

As temperatures rise, droughts deepen, and seas encroach on settlements, more areas of the planet become uninhabitable. One result of these trends is climate-driven migration. Since 2008, an average of 26.4 million people have been displaced annually by hazards and disasters; sudden-onset weather-related events account for one-third of those displacements.[1] Estimates suggest that by 2050, as many as 200 million people will be forcibly relocated due to climate-change-related impacts; most of those migrations will occur in three subregions: sub-Saharan Africa, South Asia, and Latin America.[2] We already know that most of those migrants act independently. That is, relocation has and will continue to occur ad hoc and largely without the assistance of government to coordinate new settlements, facilitate access to services, or otherwise improve the odds of a beneficial outcome for migrants or for society.

Managed retreat is a concept that suggests a more prominent role for government. Defined as "the strategic relocation of structures or abandonment of land to manage natural hazard risk," it refers to a suite of tools a government can use to position migration within the context of climate adaptation planning.[3] Doing so successfully can not only reduce harm to migrants but also powerfully expand the scope of benefits to include society at large. Situating migration within the climate adaptation milieu also identifies it as an extension of the tools we have already explored in this text.

Faced with flooding, for example, a community has three options (figure 7.1). First, the community can *protect* itself by using tools that improve resilience. This usually involves hard engineering or nature-based solutions (see chapter 3). Second, the community can better *accommodate* rising waters by adapting. Reconfiguring the built environment to withstand threats from flooding, for example, might include adopting new building codes and investing in improved early warning/early action systems (see chapter 2). But when both of those fail to eradicate

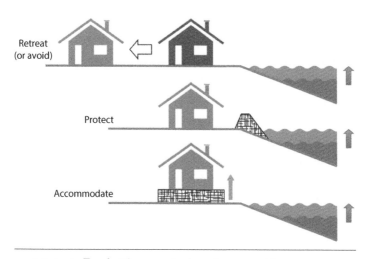

Retreat (or avoid)

Protect

Accommodate

FIGURE 7.1 Faced with a recurring hazard, communities can protect themselves, better accommodate the threat, or retreat.

Source: © 2009 GTZ

the threat, a third option exists: *retreat*. Leaving a high-risk area potentially benefits both humans and ecosystems, as society can flourish in lower-risk areas while the natural system can potentially be restored.

Migration and displacement are adjacent phenomena; in some instances, they may be difficult to untangle. Migration generally suggests a level of intent and some freedom of choice; it is defined as "the preemptive movement of people and property away from areas experiencing severe impacts."[4] By comparison, the term *displacement* generally suggests a relative lack of choice, as when people flee unsafe conditions. Both have been historically characterized by an initial shock to the household, followed by new capital accumulation strategies in a new location.

Societal mobility is, of course, not new. Push and pull factors have contributed to movement throughout history, as people have moved in search of better employment opportunities, religious freedom, improved access to land for farming, and economic stability. They have fled repressive political regimes, pervasive drought, local violence, and poverty. Most of those movements have been autonomous and uncoordinated; that is, individuals and families make their own decisions about whether to leave, where to go, and how to travel. One way to conceptualize managed retreat is across two axes, one that measures agency—who decides to move, the resident or the government—and one that measures beneficiaries—only the residents themselves or broader society. The result is a quadrant with four categories of managed retreat that establish themes for elaboration in this chapter (figure 7.2).

Migration can also be categorized as internal, with people relocating within the same country, or as international. When people relocate within a country—most commonly by leaving rural areas in favor of urban ones—their migration is difficult to manage,

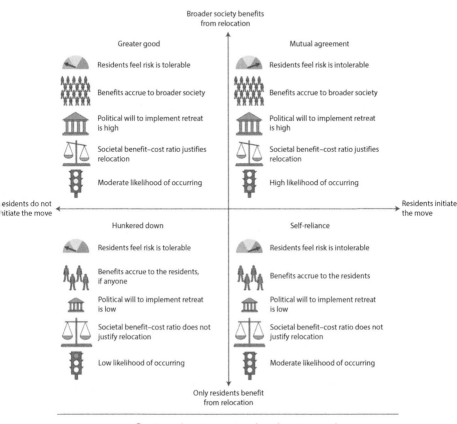

FIGURE 7.2 Sorting relocation actions by who initiates the move and who stands to benefit from it, four types of managed retreat can be identified.

Source: Image adapted courtesy of Miyuki Hino.

stop, or track. International migration is easier to measure—in 2019, an estimated 272 million people migrated internationally—and also easier to block through national border policies.[5] To date, international law does not recognize the category of "climate refugees," so we cannot accurately pinpoint how powerfully these changes to climate function as a driver of movement.

But whether an individual is moving by choice or out of necessity, and whether those movements are domestic or international, there is often an implicit connection to climate change. A threat multiplier, climate change amplifies existing risk that may stem from political conflict and resource availability. For example, as climate change disrupts historically reliable rainfall patterns, food security weakens (see chapter 5); these pressures combine with perceptions of opportunity in cities to contribute to urbanization flows.

RELOCATION: FROM WHERE?

Why do people leave? Put simply, people move when their basic needs cannot be met on-site. Rural, resource-based livelihoods have been the most deeply affected by migration through history, as people in those settings rely on a narrow range of environmental conditions for survival.[6] Departures align with variations in rainfall amounts, changes in exposure to known hazards, perceptions of food insecurity, and other natural resource allocation characteristics. Increasingly, coastal areas and other high-risk zones are also likely launch points. Sometimes these departures are understood to be short-term and cyclical in nature, with migrants returning home as seasonal conditions improve; other times the departures are considered—or become—permanent. As an economic strategy, migration has long been seen as a viable tool for addressing temporary income gaps; it functions as an informal insurance strategy. Migrants often leave family behind, establishing a system of remittances to link a future income stream with home.

But even before departing, unequal conditions exist. Only those who are able to travel—which requires at least some physical health and always comes with significant costs—can even consider migrating. Women, children, elderly persons, and those with disabilities tend to be left behind, while able-bodied men seek their fortunes elsewhere. Migrants with marketable skills, some education, and the ability to succeed in a foreign environment fare the best. Those left behind then face a weakened workforce. Even when migrants return, over time these repeated relocations have been shown to have detrimental effects on the accumulation of human capital. Each time, "displacement disrupts all aspects of people's lives, breaking social, cultural and economic networks that are critical to sustaining livelihoods, income, land ownership, and household income."[7] Thus, even when we consider migration to be voluntary and temporary, the practice is perilous (figure 7.3).

But of course, not all migration is entirely voluntary. When living conditions become so inconsistent with basic human needs—during prolonged drought or civil war, for example—people may respond to a range of incentive structures and coercive instruments as governments try to move vulnerable populations out of harm's way. One such instrument—the buyout—is discussed in detail later in this chapter. Other, less direct "push" tools include bans on new buildings through zoning laws and the withdrawal of government financial support from home insurance. At the far end of the spectrum, and still used only in isolated cases, is government-mandated resettlement. In the United States, just two communities—Newtok, Alaska, and Isle de Jean Charles, Louisiana—have forcibly moved their entire populations to pre-identified locations. Other countries have done more to engineer the location of settlements in response to emerging risks. But even when the government is trying to facilitate relocation, migratory patterns are still largely uncoordinated. Researchers note that existing policies are "typically ad hoc and focused on

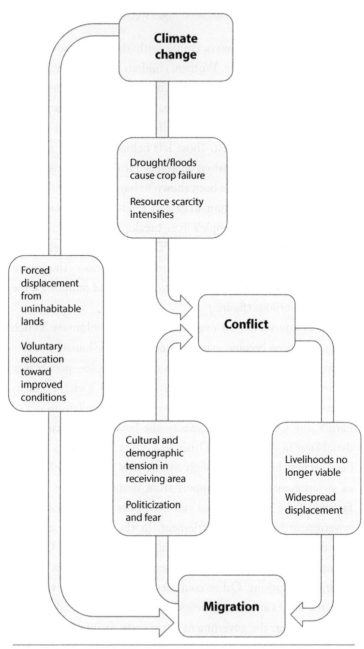

FIGURE 7.3 Climate change and resource conflict are two powerful drivers of migration, and they interact to produce amplified global pressures.

Source: Author.

risk reduction in isolation from broader societal goals."[8] As climate change impacts intensify, this represents a critical missed opportunity.

The case of Small Island Developing States (SIDS) is illuminating. Captivating the attention of the public, islanders in some of these countries have repeatedly begged for assistance as seas rise and the viability of life there is threatened.[9] Kiribati, a chain of thirty-three atolls and reef islands with a permanent population of only 110,000, has become something of an iconic face for the issue. So prominent is the issue that "in Kiribati, adaptation thinking informs political decision-making on all scales."[10] Managed retreat in Kiribati is more than an abstract notion, and the government has gone as far as purchasing land in faraway Fiji to encourage "migration with dignity" for its residents.

RELOCATION: TO WHERE?

Even when governments use a heavy hand to push people to relocate, they often fail to consider communities of destination. Where do people go? Some remain in their countries or regions of origin, either by choice or because emigration policies limit their international movements. These internal migrants numbered 41.3 million in 2018, the highest number ever recorded.[11] While internal migration is potentially less disruptive than international migration—generally, staying in the country means migrants will at least have language and culture in common with their destination community—it still creates challenging circumstances for new residents. Case studies in Belen, Peru, and Rweru, Rwanda, show that even when the government has guided resettlement, locals may find their new community to be culturally foreign; without familiar livelihood elements, some families have illegally returned to their high-risk hometowns.

Urbanization—a defining feature of global development—is, of course, itself a form of internal migration. Happening across the globe on a massive scale and accelerated by climate change, the trend creates new social challenges in cities (see chapter 4). For some, the arrival of uneducated, unskilled farm laborers seeking nutrition and safety has meant the rapid establishment of slums. In many developing regions of the world, intensifying internal migration presents a looming threat to fragile social systems.[12] International migration includes the same challenges of social dislocation as internal migration but adds layers of political and cultural complexity. Receiving communities may see incoming migrants as a potential economic drain or as a threat to national security. In many wealthier countries, today's dominant narrative describes migrants as a source of employment competition, disease, conflict, and crime. Most national policies that address the issue are designed to reduce inflows. Given this hostile welcome, it is perhaps no surprise that in many cases arriving migrants land in highly vulnerable locations, having traded unacceptable risk in their home community for different hazards in the receiving country. In these ways, international migration may not meaningfully resolve risk; instead, uncoordinated climate-driven migration has the potential to merely redistribute risk across a broader landscape.

But sometimes migrants do arrive in places that need labor, and despite populist rhetoric, these international immigrants have been shown repeatedly to contribute to economic productivity in the receiving country.[13] New arrivals with skills that match needs in the foreign labor market are best situated to contribute to cultural evolution. Indeed, "because of the diversity that accompanies migrant communities, migration acts as a vehicle for transfers of knowledge and technologies, and thus can help spur growth and development."[14] When urban planners anticipate migration and plan accordingly, these gains can be amplified.

Still, strategic planning for managed retreat is largely hypothetical. Researchers note that "there are few records of where people relocate after a retreat program. There are even fewer records of how they or others have fared: economically, socially or psychologically. Data to assess public perceptions of fairness and legitimacy are also notably missing."[15] Filling this knowledge gap will require closer monitoring and new longitudinal research. As a first step, integrating already robust international migration and refugee programs with emerging knowledge on climate-driven retreat will be key.

BUYOUTS

In some countries—mostly wealthy ones—buyout programs have emerged as the primary mechanism for governing at least the departure phase of managed retreat. For residents in flood zones, for example, a structured buyout can give them both the luxury of planning ahead and financial compensation for the move. For governments, paying residents to leave allows them to reclaim risk zones and restore the landscape. Thus, even when buyouts are costly, the tool potentially results in significant savings, as it represents a one-time expenditure and frees the government from years of repair and rebuilding costs associated with frequently flooded neighborhoods.

The U.S. experience with buyouts is instructive. Since 1978, FEMA and the Departments of Housing and Urban Development and Agriculture have all been authorized in varying situations to conduct federally led buyouts. Some states have created their own buyout programs as well, almost exclusively for coastal zones subjected to repeated flooding hazards. After Hurricane Sandy in 2012, an estimated $750 million was devoted to buyouts in the heavily populated Northeast.[16] But even then, when storm devastation across the region was severe, politicians struggled to build support for large-scale buyouts. As a result, even the most ambitious buyout plans became voluntary programs. At the neighborhood level, if only a portion of homeowners accept the buyout and vacate their property, residents remain; the local government then is profoundly restricted in the scope and scale of its revitalization efforts. In these scenarios, flood risk has not been meaningfully reduced.

Does this example mean that buyouts succeed in reducing risk only when they are made mandatory for an entire neighborhood? Perhaps, but at least three additional complications exist. First, much of the controversy around buyouts has focused on the amount the government offers to residents. Properties that once had a high value but that are now slated for demolition are difficult to price. Second, buyouts do nothing for renters. Only homeowners stand to benefit (or lose) from the instrument, and renters become expendable bystanders in an exchange that can have a powerful influence on their lives. Third, we mustn't forget the importance of social cohesion. Relationships and trust within communities are essential building blocks for adaptive capacity, and any kind of forced displacement program threatens the continuity of those place-based connections. Buyouts are focused exclusively on property and are not generally expected to capture social dimensions of migration. Taken alone, they do not comprise managed retreat. Indeed, a patchwork of buyouts is not conducive to either ecosystem restoration or long-term societal adaptation.

CASE STUDY: GATINEAU

After two one-hundred-year floods hit the small city of Gatineau within a two-year span, the Canadian government decided to take aggressive action. It adopted a new policy, focused on the neighborhood closest to the confluence of the Ottawa and Gatineau Rivers, that requires residents to leave if the damage from the latest flood exceeds 50 percent of the value of their home. Technically structured as a partial buyout, the program caps compensatory payments at $250,000 per home regardless of the home's value. Canada set the stage well ahead of this newly announced policy. The country regularly warns its citizens that living in a high-risk zone will mean little government support when expected damage occurs. Quebec, Canada's largest province, has gone even further,

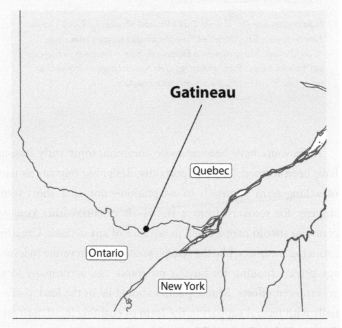

BOX FIGURE 7.1 Locator map of Gatineau, Canada.

prohibiting construction of new homes in floodplains and limiting the total amount it will pay to rebuild flooded homes anywhere. Little attention has been paid to destinations for retreating homeowners, and they are free to relocate wherever they choose. With only some homes meeting the criteria for demolition, in 2020 the neighborhood was in the midst of a massive transition. Notably, residents of Gatineau are largely supportive of the new policy; observers note that Canadians may have a natural tendency to prioritize the collective good over their own private property, a cultural feature that differentiates them from their U.S. neighbors to the south. Thus, agencies in other settings that borrow Canada's approach to managed retreat in Gatineau may confront different political and financial obstacles.

*Christopher Flavelle, "Canada Tries a Forceful Message for Flood Victims: Live Someplace Else," *New York Times*, September 10, 2019; John Carey, "Core Concept: Managed Retreat Increasingly Seen as Necessary in Response to Climate Change's Fury," *Proceedings of the National Academy of Sciences* 117, no. 24 (2020): 13182–13185, doi:10.1073/pnas.2008198117.

As buyouts have become more common, some early lessons have been learned. Most importantly, designing buyouts as part of a long-term approach to adaptation—not as a short-term strategy for recovery from a storm—is essential. This kind of planning should happen well in advance of any disaster. Creative financing measures, like the redeployment of tax revenue to leverage private funding for buyout programs, can accompany local government efforts. Municipalities should be in the lead, potentially with federal support in the form of enabling legislation and finance. Those local leaders should use data to identify priority acquisition zones and work closely with affected communities

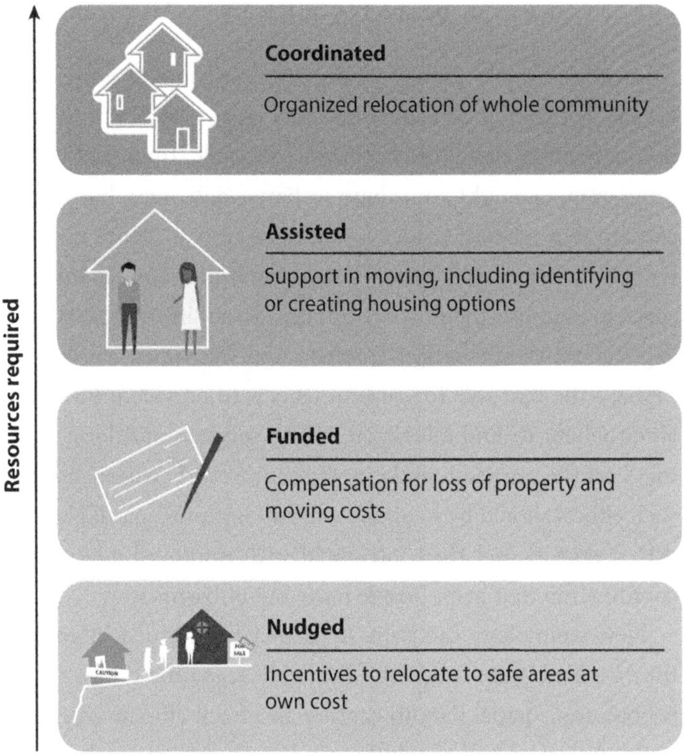

FIGURE 7.4 Governments face a continuum of possible roles in relocating at-risk communities.

Source: Illustration by Kate Marx, prepared for A. R. Siders, Miyuki Hino, and Katherine J. Mach, "The Case for Strategic and Managed Climate Retreat," *Science* 365, no. 6455 (2019): 761–763.

to chart out scenarios, time lines, and decision points. Figure 7.4 shows the variety of options governments have. We have a long way to go. Despite a consensus that "improved buyouts are an essential adaptation tool in the age of climate change,"[17] there is agreement that we are falling short: "currently, buyout programs are not structured to facilitate comprehensive and coordinated coastal adaptation strategies."[18]

WHAT'S NEXT?

Currently, virtually all migration patterns amount to unmanaged retreat. Separating climate from other drivers of migration may not be possible or, indeed, ultimately valuable. Improved buyout programming might contribute to better outcomes, but experts caution that when it comes to managed retreat, "more management is not necessarily better."[19] Relocated communities will need space to establish their own culturally driven lifestyles, and those experiences should reflect freedom and individual sovereignty. Perhaps the best role for governments is to provide information about where to find education, health services, and local housing. Lest the provision of these services be overly paternalistic, all such efforts should be rooted in a robust bottom-up stakeholder-driven process and then integrated with improved government coordination that helps people move out of harm's way.

New technology can help. A partnership formed in 2020 by the *New York Times Magazine*, ProPublica, and the Pulitzer Center has used spatial data to capture and track climate migration in the Americas.[20] Through the use of scenarios, researchers have considered the ways border policies, economic growth, and trade interact with rising GHG emissions as they explore possible future migration pathways. Anticipating migratory flows can help governments conduct necessary long-term planning. For example, some have floated the idea of building a new "climate passport" program, modeled after the Nansen passport system used during the World War I–World War II period when many stateless individuals sought safe domicile.[21] Using this approach, countries with historically high emissions might be designated to receive those with climate passports, thereby aligning the financial burden with the *polluter pays principle*.[22] One formula developed to imagine such a system would require the United States to take in as many as twenty-seven million displaced climate migrants by 2050.[23] The notion of a climate passport of course addresses only international migration, not relocation patterns within countries, but it still offers some creative fodder for consideration.

Planning for managed retreat should be as inclusive as possible. Departing stakeholders and potential receiving communities should all have an active role in organizing, funding, and executing the relocation. Experts recommend starting far ahead of time and using in situ strategies that will allow planning groups to meet; these collaborative processes will build adaptive capacity at the same time that they are developing detailed plans for a range of scenarios. Consistent with the broader notion of mainstreaming adaptation policies (see chapter 1), adding managed retreat to the toolbox of adaptation allows decision makers to utilize the movement of residents as a way to benefit society.

CONCLUSION

Managed retreat may indeed offer a promising pathway for improved adaptation outcomes—but only if those benefits are equitably allocated throughout society. Embedded injustices have undermined many promising policy efforts. The next chapter offers a deeper dive into questions of equity and justice, with a focus on how those principles can be applied to climate adaptation governance.

8

INEQUALITY AND JUSTICE

CLIMATE JUSTICE

Throughout this volume, patterns of inequity have emerged in virtually every facet of climate governance. Climate change itself is the product of an asymmetrical world. The United States and Europe have largely built their wealth through a fossil-fuel-intensive industrial revolution. They are responsible for emitting the vast majority of the GHGs that now contribute to damaging climate change impacts. Most poorer countries did not amass benefits from a transformative period of industrial development, and many are now highly vulnerable to the impacts associated with climate change. Intensifying these disparate development pathways, the countries experiencing disproportionate climate change impacts are also quite likely to suffer from pervasive capacity weaknesses—in part the result of being colonized by European powers—rendering them less able to manage climate risks. Today, the process for assessing and managing climate risks is still organized through an international technocratic model that systematically excludes the very people most at risk. Climate justice is thus about both process and outcomes and covers historical responsibility for emissions as well as current capacity to respond.

Environmental justice—a broader notion that includes climate justice—is conceptually framed as a remedy for inequities associated with access to and damages from natural resources. The study of environmental justice is, at its heart, a normative field. We begin by accepting the desirability of a world in which all people—regardless of their class, race, nationality, or ethnic origin—have equitable access to environmental benefits and contribute fairly to the costs associated with maintaining those benefits into the future.

Attention to three specific dimensions of justice can enhance efforts to build such a system in the face of multidimensional inequalities (figure 8.1). First, *distributive justice* is the allocation

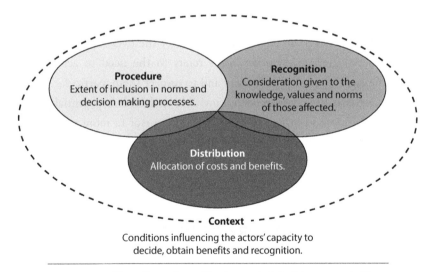

FIGURE 8.1 Three dimensions of justice can help build more equitable pathways toward climate adaptation governance.

Source: Image adapted from E. Garmendia, U. Pascual, and J. Phelps, "Environmental Justice: Instrumental for Conserving Natural Resources," BC3 Policy Briefing Series 04-15, Basque Centre for Climate Change (BC3), Bilbao, Spain, 2015.

of environmental costs and benefits across both time and space. A just approach would link the two so that people who benefit from access to a resource—say, they have cheaper energy bills as a result of burning coal—also shoulder the costs associated with it—air pollution, unsafe working conditions, and contaminated groundwater. Second, *procedural justice* refers to the process for making decisions. To be just, the process for making environmental decisions should be transparent, accessible, and fair. In the case of climate change, decisions about traditional, carbon-intensive development are often made far away from the people who are on the receiving end of a global climate crisis. The UN system is notoriously technocratic, and global policy makers have long excluded marginalized communities from their own policy development process.[1] Within countries, mirrored patterns of exclusion reflect common themes: members of ethnic and religious minorities, women, those with disabilities, and elderly persons are commonly left out of the decision-making process. Third, *recognition justice* refers to the need to accept the legitimacy of stakeholders and to respect them, particularly those that have been rendered invisible to date. It is potentially the antidote to marginalization and a precursor to more robust procedural justice. Unrecognized communities are highly vulnerable to hazard.

CONTOURS OF INEQUITY: POVERTY, GENDER, RACE, AND ETHNICITY

A closer look at the United States provides a useful example. For decades, researchers have found all three dimensions of environmental (in)justice across the country. We know that lower-income communities are more likely to be situated near

hazardous waste sites and to suffer from the toxic air and pol-
luted groundwater that accompany them.[2] Why is this so? One
explanation is that more organized, wealthier communities are
better equipped to fight the placement of that waste in their
neighborhood. Communities using this form of Not in My
Backyard (NIMBY) organizing may employ lawyers to fight
the incoming site and may draw on their access to politicians
and others with decision-making power. Another explanation
points to the tight overlap of wealth, race, and ethnicity and the
long history of institutionalized racism in the United States. For
example, "redlined" housing policies made it difficult for African
Americans to secure mortgage loans in white neighborhoods,
resulting in segregated settlement patterns, with the African
American communities situated on less valuable land closer to
toxic industrial facilities. Yet another explanation is that direct
racism itself is a driver. Chemical companies may not explicitly
value the health of minorities and immigrants, and local elected
officials may fail to adequately represent that portion of their
constituency. In this example of hazardous waste in the United
States, we may disagree on the relative strength of the various
drivers: Is it racism? Is it the inevitable result of unequal wealth?
The outcome, however, is clear: wealthier white people are better positioned to avoid the most damaging effects of industrial
waste. This is the essence of environmental injustice: the costs
and benefits are not allocated evenly across social class, race, and
gender lines.

Despite wide variations in demographic, ethnic, racial, and
income profiles across countries, many of the same patterns
of inequity exist. Majority groups tend to exclude minorities.
Lower-income households tend to have less access to power
and decision-making than higher-income households. Women
are almost universally underrepresented in political positions.

The notion of *intersectionality* perhaps captures the interconnected nature of these dynamics; it suggests that vulnerability is itself a social construction that reinforces these layers of unequal access to resources and power. In the context of climate adaptation, the relative ability to build adaptive capacity is a useful organizing principle.

One commonality bears a closer look here: gender inequities exist in every country in the world. Women are more likely to die in disasters, an outcome that is attributed to the various ways customary gender roles create patterns of vulnerability.[3] With lower levels of education, women have reduced access to timely information about impending risk. With lower rates of formalized land tenure, women are more easily displaced. With inconsistent access to health care, women are at a disadvantage even before the disaster arrives. Again, we see how root drivers of inequity that predate climate hazards are amplified by proximate threats; simultaneously, barriers to accumulation of adaptive capacity persist.

These same patterns weave through the myriad ways ethnic and racial minorities face institutionalized discrimination in many countries, rendering them more vulnerable to risk. Unjust treatment of indigenous groups by colonizing forces, including forcible removal from newly protected lands and social marginalization, has occurred throughout history and across continents. Exclusion from social benefits like education, health care, and housing has created durable gulfs between the haves and the have-nots. So stark is this divide that Desmond Tutu has described the social dynamics in climate change governance as "adaptation apartheid."[4] Low levels of social mobility—the ability of an individual born into poverty to become less poor—create stubborn class divides.

SOCIAL PROTECTION, KNOWLEDGE, AND INSTITUTIONS

Investing in social safety nets and providing equitable access to education, health care, and land tenure are obvious ways to reduce vulnerability and build adaptive capacity through government action. These tools are essential remedies for injustice, and democratic, representative governments are arguably best equipped to advance them. But building equitable systems where all residents have access to resources and power often entails institutional reform, even in egalitarian societies.

Part of the challenge is the tendency to narrowly frame the concept of *knowledge*. Over time, global institutions have embraced and legitimized a form of western environmental science that prioritizes technocratic solutions, "cutting out other ways of knowing in the process."[5] These sciences tend to define vulnerability as a static condition and do not interrogate its root causes. National-level NAP documents often manifest this framing: they define emerging climate risks, map them over existing vulnerabilities, and go on to prioritize technical infrastructure solutions.[6] Even the provision of climate services—for example, weather forecasts delivered to rural farmers—is frequently framed as top-down and linear, "cementing hierarchies of knowledge production."[7] A more equitable process would engage vulnerable communities in decision-making and service delivery from the start. It would also devote some attention to identifying root drivers of local vulnerability and considering more transformative pathways.

Global agencies and nongovernmental organizations (NGOs) have gotten the message. But systemic conceptions of knowledge are hard to shake, and new paradigms come up against entrenched institutions as "the claims made by international development actors to promote participatory and 'good governance' spaces are [still] too technocratic to empower the local groups most at risk."[8] Practitioners across the developing world have encountered these frustrating discursive barriers. In Nepal, communities were invited to participate in stakeholder meetings, but they were mostly attended by development experts and government actors; after all, those experts were being paid to devote a workday to the conversation and were already steeped in the language of adaptation.[9] Similarly, in Uganda, researchers found that participants were unable to contribute meaningfully to an adaptation workshop because draft documents were not shared ahead of time, invitations came at the last minute, and local officials were excluded.[10] These examples offer a glimpse of the layered hurdles that make equitable processes so challenging to build and sustain.

Repositioning traditional forms of knowledge might offer an on-ramp for excluded stakeholders. Too often "state institutions insist on defining traditional knowledge as a knowledge not touched by time," a romantic if inaccurate notion.[11] An updated definition should capture the dynamic ways traditional knowledge already interacts with western scientific knowledge. Valuing local knowledge doesn't mean abandoning formal education. Both are essential and can be layered for the best results. NbS are often cited for their usefulness in just this way; they integrate western engineering with the mutually beneficial relationship that already exists between humans and the natural world (see chapter 3).

FINANCING ADAPTATION

Comprehensive coverage of financial tools relevant for climate adaptation is beyond the scope of this volume. We do, however, address the topic here in a fairly limited way because one of the most visible ways to combat inequities in the climate change adaptation arena is through the delivery of financial resources. In line with the *polluter pays principle*, countries known to be responsible for emitting the majority of GHGs on their pathway to wealth should commit a portion of those gains to helping poorer countries now grappling with impacts from those emissions. If conducted on a grand scale, such transfers could potentially embody an ethical mandate for *reparations*. Here we explore the institutional structure that supports, structures, and facilitates adaptation funding.

Global Institutions

During contentious *loss and damage* negotiations at the annual UN climate conferences, poorer countries have passionately urged wealthier countries to accept financial responsibility for their costly climate change adaptation needs. So far, wealthy countries—perhaps defined in the climate policy arena as those categorized as Annex I under the Kyoto Protocol—have largely resisted these pleas, although some have noted that assisting needy countries with adaptation expenses stands to benefit everyone; after all, eco-social stability leads to reduced migration, less volatile trade, and improved food security. We might do well to remember that altruism and greed are not mutually exclusive: "Social responsibility is a moral imperative—it is also not bad for the self-interested powerful."[12] Indeed, since 1992, when the UNFCCC was established, a robust institutional structure has emerged at the global scale to facilitate the transfer of resources through a web of loans, grants, and competitive programmatic funds. An incomplete roster of those funds includes the GEF, the Green Climate Fund, and the AF. Since those early years, these institutions have disbursed millions of dollars to support climate adaptation. Still, insufficient funds and imperfect delivery systems plague adaptation funding.

With many countries focused on climate mitigation efforts, public investments in adaptation are estimated to be only 10–20 percent of the total devoted to climate change.[13] Even within that smaller adaptation space, inefficiencies persist. One reason is systemic: many of the relevant accounts are reliant on a voluntary donation system, with countries pledging support and then committing those funds; not surprisingly, there are frequent gaps—in both time and amount—between those two phases.[14] National political shifts may also contribute to this problem, as a new leader may not feel compelled to follow through on pledges made by a predecessor. But not all global adaptation funds rely on voluntary donations. For example, the AF receives 2 percent of the burgeoning market organized under the Kyoto Protocol's Clean Development Mechanism. Since this is a relatively reliable source of funds, the AF has a steady balance sheet. Perhaps there is a lesson here: embedding adaptation funding transfers more strategically into existing programs might be preferable to relying on annual voluntary payments.

Long-standing institutions that weren't specifically created for adaptation but that have compatible missions—humanitarian aid organizations and development banks come to mind—are other sources for adaptation funding. Much of the work to reduce vulnerability has long been part of the development portfolio, for example. And when adaptation policies are mainstreamed, as recommended in this text (see chapter 1), the overlap with development may be substantial. But there are also good reasons to separate the two funding streams: Improved tracking is perhaps the most obvious one, as without clearly identifiable markers for new adaptation funding, existing development projects could simply be renamed to capture adaptation benefits without contributing any new finance. Conflating the two also eliminates opportunities to flag, and potentially discontinue, development pathways that are now recognized as maladaptive. The World Bank has "climate-proofed" its development accounts, potentially addressing this concern, but national governments, private investors, and NGOs continue to push for more transparent tracking of adaptation funding.[15]

With so many overlapping and intertwined financial institutions contributing to adaptation, two sectors have emerged as the primary recipients. Infrastructure, with an estimated 40 percent of total adaptation funding, and agriculture, with an estimated 32 percent, have been favored by global finance.[16] Most of that money is released to support individual projects, and, therefore, these finance flows reflect an incremental approach to adaptation rather than a systemic transformation.[17]

But what about less tangible adaptation needs like reducing vulnerability for exposed communities? Part of the problem, we know, is that capacity begets capacity. That is, poorly equipped national governments struggle to mount competitive applications for available funds, come up with sufficient matching funds if required, and make efficient use of new funding intended, perhaps ironically, to improve their adaptive capacity. Many such under-resourced countries rely on outside experts to help them navigate the complex global finance machine. For example, tiny Tuvalu— one of the highly vulnerable SIDS facing coastal inundation— spends fully 12 percent of its total gross domestic product (GDP) on external experts and consultants.[18] Those inefficiencies not only bolster the power of highly educated countries that produce consultants but also fail to contribute to capacity building in the countries where it is needed the most.

The Private Sector

Outside of the global public climate finance milieu, other sources of funding for adaptation have emerged. The private sector is potentially the greatest source of untapped financial might, but just as the public sector has been focused on climate mitigation efforts, the private sector has also been slow to embrace adaptation opportunities. Barriers to increasing private investment in adaptation include scientific uncertainty, mismatched time horizons, and a seeming lack of tangible investment opportunities.

This is not to say progress has not been made. Indeed, the notion of resilience has permeated private investment firms, with approximately 70 percent of such organizations now screening the investments in their portfolios for potential risk from the physical effects of climate change and climate regulations.[19] But confusion about the term *resilience* persists. In part, this puzzlement stems from a failure to differentiate between making resilient investments, in which resilience is a feature of the investment opportunity, and investing in resilience, in which resilience is the product. Screening that identifies climate-resilient investment opportunities is surely a marker of the growing awareness of climate risk, but it does not amount to adaptation funding.

Yet, adaptation investment opportunities exist. Traditional investment models that use data to identify a promising company and focus capital flows there can potentially support businesses that provide products like water infrastructure, drought-resistant seeds, drip irrigation, and EWSs; additionally, firms that provide data, technological tools, and climate screening services are targets for adaptation investments. But popular investment models struggle to make quantitative sense of private investments that yield public benefits, rendering adaptation a systemically awkward fit.[20]

NbS have been particularly unsuccessful in attracting private investments. Unlike hard-infrastructure investments, which may be more easily measured, NbS "cannot be capitalized by any one part or organization. They create externalities that impact on many different groups, resulting in a problem of ownership."[21] This lack of investment is considered a major barrier to the implementation of such tools, most of which currently rely entirely on public funding sources. It's likely that research on the long-term performance of such projects will make it possible to more accurately measure their return on investment (ROI)—thus helping to attract future investors.

Blended Finance

Complex financial instruments are another type of funding for adaptation. Creative ways to blend public funding—from the international sources described earlier and national budget allocations—with private investment have bubbled up around the world (figure 8.2). For example, environmental impact bonds, first implemented in Washington, DC, as part of broader green-infrastructure investment, allow investors to get payments based on pre-defined performance like reduced runoff. Another example is catastrophe bonds: investors receive above-market returns when a prespecified catastrophe (identified through an index) does not occur. A third example is the creation of intermediary institutions that package financing from disparate sources. Latin America's Emergency Liquidity Facility fits that description; it was established in 2004 to support local microfinance institutions—essential providers of emergency funds following disasters—through the Inter-American Development Bank. Insurance at all levels of society is yet another form of finance; for a more detailed look

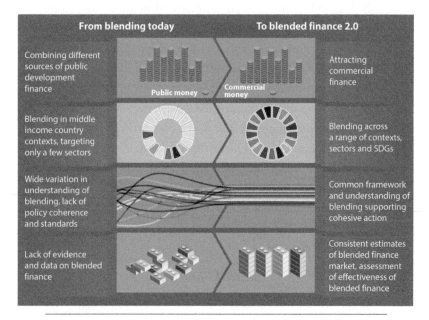

FIGURE 8.2 Blended financial tools offer potentially transformative ways to enhance the funding flows that support climate adaptation.

Source: Image: © OECD

at the role of insurance in climate adaptation, including the way catastrophe and resilience bonds can support insurance companies in covering climate risk, see chapter 6.

In the end, different kinds of finance are appropriate over different time frames and for different goals, as "how adaptation is defined determines to a large extent what and who is and is not addressed by adaptation funding."[22] Funds that are used to revitalize cities in order to reduce exposure to climate risks are distinct from funds that support recovery after a disaster. Insurance and other predisaster risk finance mechanisms are not appropriate for slow-onset impacts, and competitive project-based funding is not particularly useful for immediate postdisaster needs. Standardized and broadly embraced guidelines are sorely

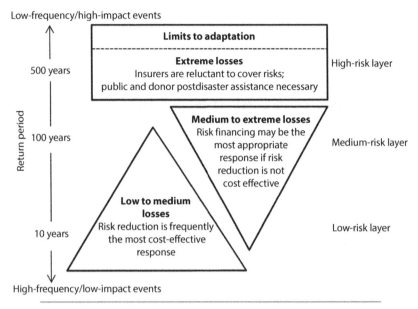

FIGURE 8.3 Different kinds of risk align with different kinds of funding to best support climate adaptation.

Source: Image adapted courtesy of Joanne Linnerooth-Bayer.

needed and suggest an important role for collaborative work that combines normative dimensions of the human rights conversation with adaptation science (figure 8.3).

CONCLUSION

It is perhaps in this broader nexus, where development meets climate science and where normative thinking meets pragmatism, that the future of climate change adaptation is most promising. Moving beyond finance to consider solutions aligned with moral principles and inspired by legal strategies, we turn now to some final thoughts.

9

SYNERGIES AND BEST PRACTICES

A s profiled in this volume, climate change adaptation governance, while still nascent, is rapidly growing. This book has provided an overview of the most salient dynamics, tools, and strategies currently guiding decision-making. We know that trade-offs are pervasive. Popular measures like infrastructure offer some protection from climate change impacts but suffer from high costs and longer-term path dependency concerns. NbS offer access points for a wider range of stakeholders at a lower cost but have not yet succeeded in attracting private investment. Agriculture policy reform can enhance the resilience of farmers but may not result in enduring improvements to food security. Zoning may address coastal risk but simultaneously may contribute to inequitable settlement patterns. Protected areas can buffer communities and conserve habitat but have a history of brutal exclusion. Insurance at all scales has a central position in adaptation, but when governments subsidize those policies, moral hazards abound. And managed retreat may be inevitable for areas facing the highest risk, but disruptions to social cohesion threaten adaptive capacity. Through it all, environmental justice remains elusive yet critical. Navigating these trade-offs in local settings is challenging. Here we conclude with some overarching thoughts about key takeaways and next steps.

REFRAMING RISK AND VULNERABILITY

This book began with an overview of key terms, and having now arrived at the end, two stand out as particularly critical: risk and vulnerability. Discourse can be powerful. The meaning held within those words drives the way we frame and understand adaptation needs and simultaneously defines and limits the responses we consider. Risk, often defined through quantitative modeling, too often envisions physical hazards interacting with static communities and high-value infrastructure; the myriad and dynamic ways vulnerability and risk interact are ignored. Human behavior, which is at least partially driven by perceptions of risk, can itself profoundly alter the risk landscape. Further, since we know that "all formal climate change risk assessments are structured by underlying values and normative goals that are sometimes explicit and sometimes hidden," we must dig deeper.[1] Identifying implicit assumptions—and, indeed, interrogating them—can begin a process in which we reimagine what risk means.

Labeling certain ethnic groups as highly vulnerable may lead to the better delivery of governmental support systems but may also contribute to perceptions that those groups are somehow inferior. Since we know that those same vulnerable groups may well have unique insights that can contribute to a meaningful adaptation strategy, identifying them as sources of valuable knowledge might shift the frame. For example, in Tanzania, the national government describes the pastoral Maasai tribe "as environmental destroyers as well as the most vulnerable people in the face of climate change, [but] the NGOs representing them argue that they are masters of adaptation instead."[2] Repositioning the tribe members as actors with agency, participating in a uniquely adaptable lifestyle, could forge new pathways toward inclusion. Incorporating indigenous knowledge is one way to reframe what vulnerability means in practice, and this is beginning to happen.

The Vulnerability and Risk Assessment (VRA) designed by Oxfam is one example of a system designed specifically to better include stakeholders outside the norm.[3]

Attention to language is not just an exercise in semantics. Case studies throughout this book have shown linkages between discourse and action, reminding us that the way we understand the world is manifest in the words we use to describe it, and those framings drive our collective behavior. Reconsidering what *risk* and *vulnerability* signify can be transformational. One way to capture the power of language is through an internationally structured articulation of shared principles.

WHAT'S NEXT?

Governing Principles

Developing a set of shared principles to guide climate change adaptation across disparate national contexts can support improved coherency and potentially more equitable outcomes. The notion is not new. Global efforts to identify, negotiate, and record a set of principles have long galvanized disparate agencies in culturally diverse settings to pursue locally appropriate policies that interpret shared values. Within the UN system, shared principles on human rights, labor, international law, and many other areas have been codified through an arduous process of mutual learning. Civil society has also been instrumental in framing shared principles for critical issues; establishing norms and values to guide climate change adaptation should also happen under this kind of big tent. For example, the Bali Principles of Climate Justice, released in 2002, begins by identifying patterns of inequity in climate change and situating the challenge of adaptation as a local one. Since then, a variety of efforts has sought to further develop connections among environmental justice and human rights, often incorporating key principles like *polluter pays*—asserting that those who emit pollution should pay for its remediation—and *common but differentiated responsibility*—noting that climate change is a shared problem but that countries bear different levels of responsibility for causing it.

Establishing a shared normative stance toward climate justice (figure 9.1) might lead planners toward "a justice perspective . . . [that] asks questions about who is able to adapt and under what circumstances; whether, in particular, adaptation is backed by a collective commitment and some notion of fairness, or is rather an ad-hoc process initiated by those with the ability and resources to do so."[4] But statements of principle lack teeth. They may play a critical role in solidifying universal values that governments can choose to codify in national laws, but they do not, in themselves, provide redress from harm. They rely on voluntary uptake.

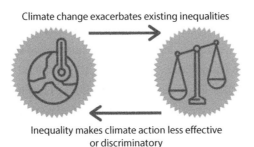

Climate change exacerbates existing inequalities

Inequality makes climate action less effective
or discriminatory

FIGURE 9.1 Climate change can amplify existing inequalities,
and those same asymmetries have an adverse effect
on the effectiveness of adaptation efforts.

Source: Infographic from "Creative Carbon Scotland's Guide to Climate Justice,"
June 25, 2020. © 2020 Creative Carbon Scotland, www.creativecarbonscotland.com.
Used with permission..

Litigation

Litigation is another approach. National legal systems have begun, only recently, to consider legal liability for climate change. The field is informed in part by the rapid growth of attribution science, a set of technical tools that measure the relative role of global climate change in driving localized disasters; for example, scientists have determined that the deadly 2019 heat waves that marinated most of western Europe in record-breaking warmth were twice as likely to occur due to anthropogenic climate change.[5] These data open the possibility for, say, a family who lost someone to heat stroke to sue a high-emitting power company for harm.

Climate litigation initially began in the United States, drawing on strategies modeled after the precedent of tobacco lawsuits, and has now spread to Asia, Europe, and other corners of the world.[6] A number of active lawsuits have elevated the issue in the courts, building case law rooted in traditional findings of liability for failure to meet the relevant standards of care in order to establish new precedent for holding fossil-fuel companies—and the governments that support them—accountable for harm that results from climate change. In a landmark 2019 decision, the Philippines Commission on Human Rights found that fossil-fuel companies likely violated the rights of citizens. The decision didn't come with penalties, as the commission does not have punitive authority, but it opens the door for future litigation and is an early signal that principles of human rights apply to energy policy. Future fees and penalties associated with climate lawsuits would likely contribute to financing for adaptation, reinforcing the connections among anthropogenic emissions, human harm, and redress.

FINAL THOUGHTS

The adaptation challenge is immense, daunting, and pressing. Although climate change is a global process, adapting to its impacts is fundamentally a local endeavor. Today, the tide of progress is swelling. The SDGs, the NDCs that galvanize individual countries to act transparently in accordance with the PA, and heightened awareness of climate-driven disasters are all promising signs. Of course, the undercurrent is insidiously strong. Wealthy actors with political power resist transformational change. Path dependency can undermine even motivated decision makers. And a scarcity of resources—including funding, technical support, and political will—bedevils implementation. But the very existence of a book like this signifies growing awareness, and that's a deeply encouraging first step. Empowering communities to understand the risks they face, build adaptive capacity, and deploy equitable solutions is the only way forward for humanity.

GLOSSARY

ADAPTATION: The process of adjustment to actual or expected climate change and its effects. In human systems, adaptation seeks to moderate or avoid harm or to maximize beneficial opportunities.

ADAPTIVE CAPACITY: The ability of a (human) system to adjust to climate change, to moderate potential damages, to take advantage of opportunities, or to cope with the consequences.

BASIS RISK: The misalignment between insurance claims and losses; it can be positive when a payout is triggered even though the disaster has not led to correlated losses, or negative when losses are extreme but payouts are insufficient.

CLIMATE RISK MANAGEMENT (CRM): A step in the adaptive policy-making process that connects real-time risk assessment with short-term and responsive management changes.

CLIMATE SERVICES: The provision of scientific climate and weather data to assist with decision-making related to climate change.

CLIMATE-SMART AGRICULTURE (CSA): An approach to cultivation and management of livestock, fisheries, and crops that supports equitable development, food security, and resilience to climate change at all levels of society.

CO-BENEFITS: The results realized when a single action benefits both climate mitigation and adaptation goals.

COMMON BUT DIFFERENTIATED RESPONSIBILITY PRINCIPLE: Underlying global climate agreements, this principle recognizes that climate change is a shared problem but that countries bear different levels of responsibility for causing it and have differing capacities for adapting to it.

CORRESPONDENCE PRINCIPLE: A tenet of economics positing that beneficiaries of a good should be the same economic actors who bear the costs of managing that good and who have the authority to make decisions regarding its equitable allocation.

DISTRIBUTIVE JUSTICE: Environmental costs and benefits should be allocated across both time and space in such a way that people who benefit from access to a resource also shoulder the costs associated with its management and use.

ECOSYSTEM SERVICES: The benefits that healthy ecosystems deliver to human societies, usually free of charge, including provisioning, regulating, cultural, and support services.

EXPOSURE: The presence of people, livelihoods, species, ecosystems, environmental functions, services, resources, infrastructure, or economic, social, or cultural assets in places and settings that could be adversely affected by hazards.

FOOD SECURITY: Physical and economic access to safe, sufficient, nutritious, and affordable food.

GOVERNANCE: The processes involved in managing a social system or modifying its members' behaviors, including actions by governments, civil society, and the private sector.

HAZARD: A natural or human-induced physical event, trend, or physical impact that may cause loss of life, injury, or other health impacts as well as damage to property, infrastructure, livelihoods, service provision, ecosystems, and environmental resources.

INSTITUTIONS: Can refer to government agencies, formalized organizations, or established practices and customs.

INTENDED NATIONALLY DETERMINED CONTRIBUTIONS (INDCS)/ NATIONALLY DETERMINED CONTRIBUTIONS (NDCS): Submissions by each country under the Paris Climate Change Agreement (2015) describing domestic mitigation and adaptation objectives. When the PA went into force, the word *intended* was dropped, and the NDCs became self-defined binding pledges.

MAINSTREAMING: The integration of adaptation policy elements into existing development, emergency management, social services, or other sectoral policy bundles.

MALADAPTATION: Climate adaptation actions that may increase risk, increase vulnerability, or contribute to diminished social welfare now or in the future.

NATIONAL ADAPTATION PLANS (NAPS): National plans encouraged under the Cancun Adaptation Framework (2010), to replace, improve and update earlier NAPAs.

NATIONAL ADAPTATION PROGRAMS OF ACTION (NAPAS): National plans encouraged under the Kyoto Protocol (1997) to identify domestic activities to be carried out in response to immediate climate threats.

NO-REGRETS POLICIES: Broadly popular measures that have few apparent trade-offs.

POLLUTER PAYS PRINCIPLE: An entity responsible for causing damage through pollution should be held responsible for remediation of those costs.

PRECAUTIONARY PRINCIPLE: In the context of scientific uncertainty, action is warranted when the potential consequences of inaction are dire.

PROCEDURAL JUSTICE: Decisions regarding resource allocation should be made using a transparent, accessible, and fair process that includes participation by relevant stakeholder groups.

RECOGNITION JUSTICE: Decision-making should explicitly acknowledge the legitimacy of all groups of stakeholders, particularly those that have been rendered invisible to date.

REPARATIONS: In the climate change context, the transfers of wealth from historically high-emitting countries to poorer countries now grappling with impacts from those emissions.

RESILIENCE: The capacity of social, economic, and environmental systems to cope with a hazardous event, trend, or disturbance by responding or reorganizing in ways that maintain their essential function, identity, and structure while also building the capacity for adaptation, learning, and transformation.

RISK: A function of dynamic interactions among hazard, exposure, and vulnerability to climate change impacts.

RISK-REFLECTIVE PRICING: An insurance pricing system that assigns higher premiums to policyholders in riskier settings.

SUBSIDIARITY PRINCIPLE: A tenet of economics positing that decisions should be made by the decision unit or layer of government at the lowest level of aggregation possible.

URBAN HEAT ISLAND: The phenomenon that occurs in densely populated cities as heat is absorbed by concrete, contributing to higher temperatures in the city than in surrounding areas.

VULNERABILITY: The propensity or predisposition to be adversely affected; it encompasses a variety of concepts and elements, including sensitivity or susceptibility to harm and lack of capacity to cope and adapt.

WICKED PROBLEM: A problem that defies a clear solution; solving it means balancing trade-offs and accepting some harm in pursuit of some good.

NOTES

INTRODUCTION

1. Intergovernmental Panel on Climate Change, *Fifth Assessment Report of the Intergovernmental Panel on Climate Change* (Cambridge: Cambridge University Press, 2014). All citations in the text, unless otherwise noted, are to this report.

2. Jakob Zscheischler, Seth Westra, Bart Hurk, Sonia Seneviratne, Philip Ward, Andy Pitman, Amir AghaKouchak, David Bresch, Michael Leonard, Thomas Wahl, and Xuebin Zhang, "Future Climate Risk from Compound Events," *Natural Climate Change* 8 (2018): 470, doi:10.1038/s41558-018-0156-3.

3. Ilan Kelman, "Lost for Words Amongst Disaster Risk Science Vocabulary?," *International Journal of Disaster Risk Science* 9, no. 3 (2018): 282, doi:10.1007/s13753-018-0188-3.

4. Andrew Jordan Wilson and Ben Orlove, "What Do We Mean When We Say Climate Change Is Urgent?" (working paper 1, Center for Research on Environmental Decisions, New York, 2019).

5. Kelman, "Lost for Words," 284.

6. Kelman, "Lost for Words," 281–291.

7. W. Neil Adger, Iain Brown, and Swenja Surminski, "Advances in Risk Assessment for Climate Change Adaptation Policy," *Philosophical Transactions of the Royal Society A: Mathematical, Physical and Engineering Sciences* 376, no. 2121 (2018): 6, doi:10.1098/rsta.2018.0106.

8. Jonas Hein and Yvonne Kunz, "Adapting in a Carbon Pool? Politicising Climate Change at Sumatra's Oil Palm Frontier," in *A Critical Approach to Climate Change Adaptation: Discourses, Policies, and Practices*, ed. S. Klepp and L. Chavez-Rodriguez (Oxon, UK: Routledge, 2018), 162.

9. Intergovernmental Panel on Climate Change, "Summary for Policymakers," in *Climate Change 2007: Impacts, Adaptation and Vulnerability; Contribution of Working Group II to the Fourth Assessment Report of the Intergovernmental Panel on Climate Change*, ed. M. L. Parry, O. F. Canziani, J. P. Palutikof, P. J. van der Linden, and C. E. Hanson (Cambridge: Cambridge University Press, 2007), 7–22.

10. Sybille Bauriedl and Detlaf Müller-Mahn, "Conclusion: The Politics in Critical Adaptation Research," in *A Critical Approach to Climate Change Adaptation: Discourses, Policies, and Practices*, ed. S. Klepp and L. Chavez-Rodriguez (Oxon, UK: Routledge, 2018), 285.

11. Susanne Moser, Sara Meerow, James Arnott, and Emily Jack-Scott, "The Turbulent World of Resilience: Interpretations and Themes for Transdisciplinary Dialogue," *Climatic Change* 153, no. 1 (2019): 21–40, doi:10.1007/s10584-018-2358-0.

12. Christophe Béné, Derek Headey, Lawrence Haddad, and Klaus von Grebmer, "Is Resilience a Useful Concept in the Context of Food Security and Nutrition Programmes? Some Conceptual and Practical Considerations," *Food Security* 8, no. 1 (2016): 123–138, doi:10.1007/s12571-015-0526-x.

13. Silja Klepp and Libertad Chavez-Rodriguez, "Governing Climate Change: The Power of Adaptation Discourses, Policies, and Practices," in *A Critical Approach to Climate Change Adaptation: Discourses, Policies, and Practices*, ed. S. Klepp and L. Chavez-Rodriguez (Oxon, UK: Routledge, 2018), 15.

14. A. K. Magnan, E. L. F. Schipper, M. Burkett, S. Bharwani, I. Burton, S. Eriksen, F. Gemenne, J. Schaar, and G. Ziervogel, "Addressing the Risk of Maladaptation to Climate Change," *WIREs Climate Change* 7, no. 5 (2016): 646–665, doi:10.1002/wcc.409.

15. Jon Barnett, Saffron O'Neil, Steve Waller, and Sarah Rogers, "Reducing the Risk of Maladaptation in Response to Sea-Level Rise and Urban Water Scarcity," in *Successful Adaptation to Climate Change: Linking Science and Policy in a Rapidly Changing World*, ed. S. C. Moser and M. T. Boykoff (Oxon, UK: Routledge, 2013), 37–49.

1. FOUNDATIONS: SCIENCE, POLICY, AND INSTITUTIONS

1. Christoph Clar and Reinhard Steurer, "Why Popular Support Tools on Climate Change Adaptation Have Difficulties in Reaching Local Policy-Makers: Qualitative Insights from the UK and Germany," *Environmental Policy and Governance* 28, no. 3 (2018): 172–182, doi: 10.1002/eet.1802.

2. Andrew Jordan Wilson and Ben Orlove, "What Do We Mean When We Say Climate Change Is Urgent?" (working paper 1, Center for Research on Environmental Decisions, New York, 2019).

3. Clar and Steurer, "Why Popular Support Tools," 174.

4. Anthony Patt, "Climate Risk Management: Laying the Groundwork for Successful Adaptation," in *Successful Adaptation to Climate Change: Linking Science and Policy in a Rapidly Changing World*, ed. S. C. Moser and M. T. Boykoff (Oxon, UK: Routledge, 2013), 186–200.

5. United Nations, *United Nations Framework Convention on Climate Change* (1992), 9, https://unfccc.int/files/essential_background/background _publications_htmlpdf/application/pdf/conveng.pdf.

6. Christoph Clar, Andrea Prutsch, and Reinhard Steurer, "Barriers and Guidelines for Public Policies on Climate Change Adaptation: A Missed Opportunity of Scientific Knowledge-Brokerage," *Natural Resources Forum* 37, no. 1 (2013): 1–18, doi:10.1111/1477-8947.12013.

7. Filomena Pietrapertosa, Valeriy Khokhlov, Monica Salvia, and Carmelina Cosmi, "Climate Change Adaptation Policies and Plans: A Survey in 11 South East European Countries," *Renewable and Sustainable Energy Reviews* 81 (2018): 3041–3050, doi:10.1016/j.rser .2017.06.116.

8. Kai A. Konrad and Marchel Thum, "The Role of Economic Policy in Climate Change Adaptation," *CESifo Economic Studies* 60, no. 1 (2014): 49, doi:10.1093/cesifo/ift003.

9. Issah Justice Musah-Surugu, Albert Ahenkan, and Justice Nyigmah Bawole, "Too Weak to Lead: Motivation, Agenda Setting and Constraints of Local Government to Implement Decentralized Climate Change Adaptation Policy in Ghana," *Environment, Development and Sustainability* 21, no. 2 (2017): 591, doi:10.1007 /s10668-017-0049-z.

10. Anu Bask, Merja Halme, Markku Kallio, and Markku Kuula, "Consumer Preferences for Sustainability and Their Impact on Supply Chain Management: The Case of Mobile Phones," *International Journal of Physical Distribution and Logistics Management* 43, no. 5/6 (2013): 380–406, doi:10.1108/IJPDLM-03-2012-0081.

11. Daniel Puig, Anne Olhoff, Skylar Bee, Barney Dickson, and Keith Alverson, *The Adaptation Finance Gap Report* (United Nations Environment Program, 2016), 1–73.

12. Jay Koh, Emilie Mazzacurati, and Stacy Swann, "Bridging the Adaptation Gap: Approaches to Measurement of Physical Climate Risk and Examples of Investment in Climate Adaptation and Resilience" (discussion paper, Global Adaptation and Resilience Investment Working Group, New York, 2016), 2.

13. Rasmus Kløcker Larsen, Åsa Gerger Swartling, Neil Powell, Brad May, Ryan Plummer, Louise Simonsson, and Maria Osbeck, "A Framework for Facilitating Dialogue Between Policy Planners and Local Climate Change Adaptation Professionals: Cases from Sweden, Canada and Indonesia," *Environmental Science and Policy* 23 (2012): 20, doi:10.1016/j.envsci.2012.06.014.

14. Johann Dupuis and Robbert Biesbroek, "Comparing Apples and Oranges: The Dependent Variable Problem in Comparing and Evaluating Climate Change Adaptation Policies," *Global Environmental Change* 23, no. 6 (2013): 1476–1487, doi:10.1016/j.gloenvcha.2013.07.022.

15. Global Commission on Adaptation (GCA), *Adapt Now: A Global Call for Leadership on Climate Resilience* (Rotterdam: GCA, 2019).

16. Heleen L. P. Mees, Justin Dijk, Daan van Soest, Peter P. J. Driessen, Marleen H. F. M. W. van Rijswick, and Hens Runhaar, "A Method for the Deliberate and Deliberative Selection of Policy Instrument Mixes for Climate Change Adaptation," *Ecology and Society* 19, no. 2 (2014): 1–16, doi:10.5751/ES-06639-190258.

17. Robbert Biesbroek and Jeroen J. L. Candel, "Mechanisms for Policy (Dis)integration: Explaining Food Policy and Climate Change Adaptation Policy in the Netherlands," *Policy Sciences* 53, no. 1 (2020): 61–84, doi:10.1007/s11077-019-09354-2.

18. Frans Berkhout, Laurens M. Bouwer, Joanne Bayer, Maha Bouzid, Mar Cabeza, Susanne Hanger, Andries Hof, Paul Hunter, Laura Meller,

Anthony Patt, Benjamin Pfluger, Tim Rayner, Kristin Reichardt, and Astrid van Teeffelen, "European Policy Responses to Climate Change: Progress on Mainstreaming Emissions Reduction and Adaptation," *Regional Environmental Change* 15, no. 6 (2015): 955, doi:10.1007/s10113 -015-0801-6.

19. Lindsay C. Stringer, Andrew J. Dougill, Jen C. Dyer, Katharine Vincent, Florian Fritzsche, Julia Leventon, Mario Paulo Falcão, Pacal Manyakaidze, Stephen Syampungani, Philip Powell, and Gabriel Kalaba, "Advancing Climate Compatible Development: Lessons from Southern Africa," *Regional Environmental Change* 14, no. 2 (2014): 713–725, doi:10.1007/s10113-013-0533-4.

20. Larsen et al., "A Framework for Facilitating Dialogue," 12–23.

21. Mya Sherman, Lea Berrang-Ford, Shuaib Lwasa, James Ford, Didacus B. Namanya, Alejandro Llanos-Cuentes, Michelle Maillet, and Sherilee Harper, "Drawing the Line Between Adaptation and Development: A Systematic Literature Review of Planned Adaptation in Developing Countries," *WIREs Climate Change* 7, no. 5 (2016): 708, doi:10.1002/wcc.416.

2. DISASTER RISK MANAGEMENT: EARLY WARNING, EARLY ACTION

1. Hannah Ritchie and Max Roser, "Natural Disasters," Our World in Data, 2021, https://ourworldindata.org/natural-disasters.

2. Aon plc, *Weather and Climate Catastrophe Insight: 2019 Annual Report* (Chicago: Aon plc, 2020), 1.

3. National Academy of Sciences, Engineering, and Medicine, *Attribution of Extreme Weather Events in the Context of Climate Change* (Washington, DC: National Academies Press, 2016), 1.

4. United Nations Office for Disaster Risk Reduction (UNDRR), "Our Work," UNDRR, last modified 2020, https://www.undrr.org /about-undrr/our-work.

5. United Nations Office for Disaster Risk Reduction (UNDRR), *2019 Annual Report* (Geneva: United Nations, 2020), 12.

6. Greg Bankoff, "Time Is of the Essence: Disasters, Vulnerability and History," *International Journal of Mass Emergencies and Disasters* 22, no. 3 (2004): 5.

7. Federal Emergency Management Agency, *2019 National Preparedness Report* (Washington, DC: U.S. Department of Homeland Security, 2019).

8. UNDRR, *2019 Annual Report*, 13.

9. United Nations Development Program (UNDP), *Five Approaches to Build Functional Early Warning Systems* (New York: UNDP, 2018), 5.

10. Global Commission on Adaptation (GCA), *Adapt Now: A Global Call for Leadership on Climate Resilience* (Rotterdam: GCA, 2019), 5.

11. Lisa Dale, Jonathan Held, Stephanie Ratte, and Farrukh Zaman, "Anticipate, Absorb, Reshape (A2R): A Baseline Study of Climate Resilience in Developing Countries," *Consilience* 19 (2018): 1–22, doi: 10.7916/consilience.v0i19.3941.

12. Anthony Patt, "Climate Risk Management: Laying the Groundwork for Successful Adaptation," in *Successful Adaptation to Climate Change: Linking Science and Policy in a Rapidly Changing World*, ed. S. C. Moser and M. T. Boykoff (Oxon, UK: Routledge, 2013), 186–200.

13. Ben Orlove, Racheal Shwom, Ezra Markowitz, and So-Min Cheong, "Climate Decision-Making," *Annual Review of Environment and Resources* 45 (2020): 1–33, doi:10.1146/annurev-environ-012320-085130.

14. American Meteorological Society (AMS), "A Policy Statement of the American Meteorological Society," AMS, September 17, 2015, https://www.ametsoc.org/index.cfm/ams/about-ams/ams-statements /statements-of-the-ams-in-force/climate-services1/#:~:text=Climate %20services%20(CS)%20may%20be,collaboration%20between %20providers%20and%20users.

15. Jakob Zscheischler, Seth Westra, Bart Hurk, Sonia Seneviratne, Philip Ward, Andy Pitman, Amir AghaKouchak, David Bresch, Michael Leonard, Thomas Wahl, and Xuebin Zhang, "Future Climate Risk from Compound Events," *Natural Climate Change* 8 (2018): 469–477, doi:10.1038/s41558-018-0156-3.

16. Christoph Clar and Reinhard Steurer, "Why Popular Support Tools on Climate Change Adaptation Have Difficulties in Reaching Local Policy-Makers: Qualitative Insights from the UK and Germany," *Environmental Policy and Governance* 28, no. 3 (2018): 172–182, doi:10.1002 /eet.1802.

17. Clar and Steurer, "Why Popular Support Tools," 175.

18. Simon O'Dea, "Smartphone Penetration Worldwide as Share of Global Population 2016–2020," Statista, June 2021, https://www.statista.com/statistics/203734/global-smartphone-penetration-per-capita-since-2005/#statisticContainer.

19. Mohammed Rondhi, Ahmad Fatikhul Khasen, Yasuhiro Mori, and Takumi Kondo, "Assessing the Role of the Perceived Impact of Climate Change on National Adaptation Policy: The Case of Rice Farming in Indonesia," *Land* 8, no. 5 (2019): 15, doi:10.3390/land8050081.

3. THE BUILT ENVIRONMENT: INFRASTRUCTURE AND NATURE-BASED SOLUTIONS

1. World Economic Forum (WEF), *The Global Risks Report 2019* (Geneva: WEF, 2019), 56.

2. Borja G. Reguero, Michael W. Beck, David N. Bresch, Juliano Calil, and Imen Meliane, "Comparing the Cost Effectiveness of Nature-Based and Coastal Adaptation: A Case Study from the Gulf Coast of the United States," *PLoS ONE* 13, no. 4 (2018): 1–24, 2, doi:10.1371/journal.pone.0192132.

3. Rachel K. Gittman, F. Joel Fodrie, Alyssa M. Popowich, Danielle A. Keller, John F. Bruno, Carolyn A. Currin, Charles H. Peterson, and Michael F. Piehler, "Engineering Away Our Natural Defenses: An Analysis of Shoreline Hardening in the US," *Frontiers in Ecology and the Environment* 13, no. 6 (2015): 301–7, https://esajournals.onlinelibrary.wiley.com/doi/full/10.1890/150065.

4. Thu Thi Nguyen, Jamie Pittock, and Bich Huong Nguyen, "Integration of Ecosystem-Based Adaptation to Climate Change Policies in Viet Nam," *Climatic Change* 142, no. 1–2 (2017): 97–111, doi:10.1007/s10584-017-1936-x.

5. Nathalie Seddon, Alexandre Chausson, Pam Berry, Cécile A. J. Girardin, Alison Smith, and Beth Turner, "Understanding the Value and Limits of Nature-Based Solutions to Climate Change and Other Global Challenges," *Philosophical Transactions of the Royal Society B: Biological Sciences* 375, no. 1794 (2020): 1–12, 5, doi:10.1098/rstb.2019.0120.

6. Bill Millard, "Symposium: Extreme Heat: Hot Cities—Adapting to a Hotter World" (Design for Risk and Reconstruction Committee of the American Institute of Architects' New York Chapter, New York, November 12, 2015).

7. Michael Mullan, Lisa Danielson, Berenice Lasfargues, Naeeda Chrishna Morgado, and Edward Perry, *Climate-Resilient Infrastructure*, Environment Policy Paper No. 14 (Paris: Organization for Economic Cooperation and Development, 2018).

8. Andrea Prutsch, Torsten Grothmann, Sabine McCallum, Inke Schauser, and Rob Swart, eds., *Climate Change Adaptation Manual: Lessons Learned from European and Other Industrialised Countries* (Oxon, UK: Routledge, 2014), 227.

9. Nishtha Manocha and Vladan Babovic, "Planning Flood Risk Infrastructure Development Under Climate Change Uncertainty," *Procedia Engineering* 154 (2016): 1407, doi:10.1016/j.proeng.2016.07.511.

10. Roger E. Kasperson and Bonnie Ram, "Rapid Transformation of the US Electric Power System: Prospects and Impediments," in *Successful Adaptation to Climate Change: Linking Science and Policy in a Rapidly Changing World*, ed. S. C. Moser and M. T. Boykoff (Oxon, UK: Routledge, 2013), 114–131.

11. Nguyen et al., "Integration of Ecosystem-Based Adaptation."

12. Jennifer Hewett, "Scott Morrison Doubles Down on Climate Change," *Financial Review*, January 29, 2020, https://www.afr.com/politics/federal/scott-morrison-doubles-down-on-climate-change-20200129-p53vwc.

13. L. W. Sussams, W. R. Sheate, and R. P. Eales, "Green Infrastructure as a Climate Change Adaptation Policy Intervention: Muddying the Waters or Clearing a Path to a More Secure Future?," *Journal of Environmental Management* 147 (2015): 184, doi:10.1016/j.jenvman.2014.09.003.

14. Prutsch et al., *Climate Change Adaptation Manual*, 227.

15. Kueoi-Hsien Liao, Shinuo Deng, and Puay Yok Tan, "Blue-Green Infrastructure: New Frontier for Sustainable Urban Stormwater Management," in *Greening Cities: Advances in 21st Century Human Settlements*, ed. P. Y. Tan and C. Y. Jim (Singapore: Springer, 2017), 203–226.

16. J. Sörensen and T. Emilsson, "Evaluating Flood Risk Reduction by Urban Blue-Green Infrastructure Using Insurance Data," *Journal of*

Water Resources Planning and Management 145, no. 2 (2019), doi:10.1061 /(ASCE)WR.1943-5452.0001037.

17. Marthe L. Derkzen, Astrid J. A. van Teefelen, and Peter H. Verburg, "Green Infrastructure for Urban Climate Adaptation: How Do Residents' Views on Climate Impacts and Green Infrastructure Shape Adaptation Preferences?," *Landscape and Urban Planning* 157 (2017): 106–130, doi:10.1016/j.landurbplan.2016.05.027.

18. Isabelle Anguelovski, James J. T. Connolly, Hamil Pearsall, Galia Shokry, Melissa Checker, Juliana Maantay, Kenneth Gould, Tammy Lewis, Andrew Maroko, and J. Timmons Roberts, "Opinion: Why Green 'Climate Gentrification' Threatens Poor and Vulnerable Populations," *Proceedings of the National Academy of Sciences* 116, no. 52 (2019): 26139–26143 26141, doi:10.1073/pnas.1920490117.

19. Meredith Wiggins, "Eroding Paradigms: Heritage in an Age of Climate Gentrification," *Change Over Time* 8, no. 1 (2018): 122–130, doi:10.1353 /cot.2018.0006.

20. "Nature-Based Solutions," Commission on Ecosystem Management, International Union for Conservation of Nature, last modified 2020, https:// www.iucn.org/commissions/commission-ecosystem-management /our-work/nature-based-solutions.

21. Global Commission on Adaptation (GCA), *Adapt Now: A Global Call for Leadership on Climate Resilience* (Rotterdam: GCA, 2019), 5.

22. Nguyen et al., "Integration of Ecosystem-Based Adaptation."

23. Niki Frantzeskaki, Timon McPhearson, Marcus J. Collier, Dave Kendal, Harriet Bulkeley, Adina Dumitru, Claire Walsh, Kate Noble, Ernita van Wyk, Camilo Ordóñez, Cathay Oke, and László Pintér, "Nature-Based Solutions for Urban Climate Change Adaptation: Linking Science, Policy, and Practice Communities for Evidence-Based Decision-Making," *BioScience* 69, no. 6 (2019): 455–466, doi: 10.1093/biosci/biz042.

24. Seddon et al., "Understanding the Value and Limits," 2.

25. Colin F. Quinn, Jennifer F. Howard, Chen Chen, Joyce E. Coffee, Carlos E. Quintela, Britt A. Parker, and Joel B. Smith, "Adaptation and Poverty Reduction in Mozambique: An Opportunity for Developing Countries to Lead," *Climate Policy* 18, no. 2 (2018): 147, doi:10.1080/1469 3062.2016.1258631.

26. Nguyen et al., "Integration of Ecosystem-Based Adaptation."

27. Borja G. Reguero, Michael W. Beck, David N. Bresch, Juliano Calil, and Imen Meliane, "Comparing the Cost Effectiveness of Nature-Based and Coastal Adaptation: A Case Study from the Gulf Coast of the United States," *PLoS ONE* 13, no. 4 (2018): 10, doi:10.1371/journal .pone.0192132.

28. Seddon et al., "Understanding the Value and Limits," 8.

29. Elisabeth Hamin, Yaser Abunnasr, Max Roman Dilthey, Pamela K. Judge, Melissa A. Kenney, Paul Kirshen, Thomas C. Sheahan, Don J. De Groot, Robert L. Ryan, Brian G. McAdoo, Leonard Nurse, Jane A. Buxton, Ariana E. Sutton-Grier, Marielos Arlen Marin, and Rebecca Fricke, "Pathways to Coastal Resiliency: The Adaptive Gradients Framework," *Sustainability* 10, no. 8 (2018): 1–20, doi:10.3390 /su10082629.

4. URBAN PLANNING FOR CLIMATE ADAPTATION

1. "Adaptation in EU Sectors: Urban," Climate ADAPT, last modified 2020, https://climate-adapt.eea.europa.eu/eu-adaptation-policy/sector -policies/urban.

2. Global Commission on Adaptation (GCA), *Adapt Now: A Global Call for Leadership on Climate Resilience* (Rotterdam: GCA, 2019), 39.

3. Faustin Tirwirukwa Kalabamu, "Land Tenure Reforms and Persistence of Land Conflicts in Sub-Saharan Africa: The Case of Botswana," *Land Use Policy* 81 (2019): 339, doi:10.1016/j.landusepol.2018.11.002.

4. Osman Balaban and Meltem Şenol Balaban, "Adaptation to Climate Change: Barriers in the Turkish Local Context," *TeMA: Journal of Land Use, Mobility and Environment* (2015): 7–22, doi:10.6092/1970-9870/3650.

5. Malcolm Araos, Lea Berrang-Ford, James D. Ford, Stephanie E. Austin, Robert Biesbroek, and Alexandra Lesnikowski, "Climate Change Adaptation Planning in Large Cities: A Systematic Global Assessment," *Environmental Science and Policy* 66 (2016): 375–382, doi:10.1016/j. envsci.2016.06.009.

6. Walter Leal Filho, Abdul-Lateef Balogun, Desalegn Yayeh Ayal, E. Matthew Bethurem, Miriam Murambadoro, Julia Mambo, Habitamu Taddese, Gebrekidan Worku Tefera, Gustavo J. Nagy, Hubert Fudjumdjum, and Paschal Mugabe, "Strengthening Climate Change Adaptation Capacity in Africa: Case Studies from Six Major African Cities

and Policy Implications," *Environmental Science and Policy* 86 (2018): 29–37, doi:10.1016/j.envsci.2018.05.004.

7. Alex De Sherbinin, Andrew Schiller, and Alex Pulsipher, "The Vulnerability of Global Cities to Climate Hazards," *Environment and Urbanization* 19, no. 1 (2007): 39–64, doi:10.1177/0956247807076725.

8. Niki Frantzeskaki, Timon McPhearson, Marcus J. Collier, Dave Kendal, Harriet Bulkeley, Adina Dumitru, Claire Walsh, Kate Noble, Ernita van Wyk, Camilo Ordóñez, Cathay Oke, and László Pintér, "Nature-Based Solutions for Urban Climate Change Adaptation: Linking Science, Policy, and Practice Communities for Evidence-Based Decision-Making," *BioScience* 69, no. 6 (2019): 459, doi:10.1093/biosci/biz042.

9. Bill Millard, *Extreme Heat: Hot Cities—Adapting to a Hotter World* (New York: Design for Risk and Reconstruction Committee, American Institute of Architects New York, 2016), 3.

10. World Bank, *Doing Business 2015: Going Beyond Efficiency* (Washington, DC: World Bank, 2015), doi:10.1596/978-1-4648-0351-2, 55.

11. De Sherbinin et al., "The Vulnerability of Global Cities."

12. Malcolm Araos, Lea Berrang-Ford, James D. Ford, Stephanie E. Austin, Robbert Biesbroek, and Alexandra Lesnikowski, "Climate Change Adaptation Planning in Large Cities: A Systematic Global Assessment," *Environmental Science and Policy* 66 (December 2016): 375–382, doi:10.1016/j.envsci.2016.06.009, 377.

13. Filho et al., "Strengthening Climate Change Adaptation Capacity," 35.

5. AGRICULTURE, LAND USE, AND FOOD SECURITY

1. Stephane Hallegatte, Mook Bangalore, Laura Bonzanigo, Marianne Fay, Tamaro Kane, Ulf Narloch, Julie Rozenberg, David Treguer, and Adrien Vogt-Schilb, *Shock Waves: Managing the Impacts of Climate Change on Poverty*, Climate Change and Development Series (Washington, DC: World Bank, 2016).

2. Susan A. Crate, "Atlases of Community Change: Community Collaborative-Interactive Projects in Russia and Canada," in *A Critical Approach to Climate Change Adaptation: Discourses, Policies, and Practices*, ed. S. Klepp and L. Chavez-Rodriguez (Oxon, UK: Routledge, 2018), 242.

3. Global Commission on Adaptation (GCA), *Adapt Now: A Global Call for Leadership on Climate Resilience* (Rotterdam: GCA, 2019), 24.
4. Intergovernmental Panel on Climate Change (IPCC), *Climate Change and Land* (Cambridge: Cambridge University Press, 2019), 8.
5. IPCC, *Climate Change and Land*, 5.
6. Goytom Abraha Kahsey and Lars Gårn Hansen, "The Effect of Climate Change and Adaptation Policy on Agricultural Production in Eastern Africa," *Ecological Economics* 121 (2016): 54–64, doi:10.1016/j.ecolecon.2015.11.016.
7. Robert Zougmore, Samuel Partey, Mathieu Ouedraogomidele Omitoyin, Timothy Thomas, Augustine Ayantunde, Polly Ericksen, Mohammed Said, and Abdulai Jalloh, "Toward Climate-Smart Agriculture in West Africa: A Review of Climate Change Impacts, Adaptation Strategies and Policy Developments for the Livestock, Fishery and Crop Production Sectors," *Agriculture and Food Security* 5 (2016): 1–16, 7, doi: 10.1186/s40066-016-0075-3.
8. Mohammed Rondhi, Ahmad Fatikhul Khasen, Yasuhiro Mori, and Takumi Kondo, "Assessing the Role of the Perceived Impact of Climate Change on National Adaptation Policy: The Case of Rice Farming in Indonesia," *Land* 8, no. 5 (2019): 1–21, doi:10.3390/land8050081.
9. Kenshi Baba and Mitsuru Tanaka, "Attitudes of Farmers and Rural Area Residents Toward Climate Change Adaptation Measures: Their Preferences and Determinants of Their Attitudes," *Climate* 7, no. 5 (2019): 1–11, doi:10.3390/cli7050071.
10. Yuta J. Masuda, Brianna Castro, Ike Aggraeni, Nicholas H. Wolff, Kristie Ebi, Teevrat Garg, Edward T. Game, Jennifer Krenz, and June Spector, "How Are Healthy, Working Populations Affected by Increasing Temperatures in the Tropics? Implications for Climate Change Adaptation Policies," *Global Environmental Change* 56 (2019): 38, doi:10.1016/j.gloenvcha.2019.03.005.
11. Kahsey and Hansen, "The Effect of Climate Change and Adaptation Policy."
12. Ann Jaworski, "Encouraging Climate Adaptation Through Reform of Federal Crop Insurance Subsidies Notes," *New York University Law Review* 91, no. 6 (2016): 1684–1718.

13. Laura Bonzanigo, Dragana Bojovic, Alexandros Maziotis, and Carlo Giupponi, "Agricultural Policy Informed by Farmers' Adaptation Experience to Climate Change in Veneto, Italy," *Regional Environmental Change* 16, no. 1 (2016): 245–258, doi:10.1007/s10113-014-0750-5; Pytrik Reidsma, Frank Ewert, Alfons Oude Lansink, and Rik Leemans, "Adaptation to Climate Change and Climate Variability in European Agriculture: The Importance of Farm Level Responses," *European Journal of Agronomy* 32, no. 1 (2010): 91–102, doi:10.1016/j.eja.2009.06003.

14. World Bank, "World Development Indicators," 2020, https://datacatalog .worldbank.org/dataset/world-development-indicators.

15. Bonzanigo et al., "Agricultural Policy.".

16. World Bank, "World Development Indicators."

17. Rondhi et al., "Assessing the Role of the Perceived Impact."

18. Zougmore et al., "Toward Climate-Smart Agriculture in West Africa," 1–16.

19. International Union for the Conservation of Nature (IUCN), *Guidelines for Applying Protected Area Management Categories* (Gland, Switz.: IUCN, 2008), 2.

20. United Nations Environment Program World Conservation Monitoring Centre (UNEP-WCMC), International Union for the Conservation of Nature (IUCN), and National Geographic Society (NGS), *Protected Planet Report 2018* (Cambridge: UNEP-WCMC, IUCN, NGS, 2018).

21. Faustin Tirwirukwa Kalabamu, "Land Tenure Reforms and Persistence of Land Conflicts in Sub-Saharan Africa: The Case of Botswana," *Land Use Policy* 81 (2019): 337–345, doi:10.1016/j.landusepol.2018.11.002.

22. Hallegatte et al., *Shock Waves.*

23. Edidah L. Ampaire, Laurence Jassogne, Happy Providence, Mariola Acosta, Jennifer Twyman, Leigh Winowiecki, and Piet van Asten, "Institutional Challenges to Climate Change Adaptation: A Case Study on Policy Action Gaps in Uganda," *Environmental Science and Policy* 75 (2017): 81–90, doi:10.1016/j.envsci.2017.05.013.

24. Andrea Prutsch, Torsten Grothmann, Sabine McCallum, Inke Schauser, and Rob Swart, eds., *Climate Change Adaptation Manual: Lessons Learned from European and Other Industrialized Countries* (Oxon, UK: Routledge, 2014), 132.

6. INSURANCE AS RISK TRANSFER

1. Anthony Patt, "Climate Risk Management: Laying the Ground-work for Successful Adaptation," in *Successful Adaptation to Climate Change: Linking Science and Policy in a Rapidly Changing World*, ed. S. C. Moser and M. T. Boykoff (Oxon, UK: Routledge, 2013), 186–200.

2. InsuResilience Global Partnership (IGP), *Annual Report 2018: Working Towards a Climate-Resilient Future* (Bonn: IGP, 2018), and *Annual Report 2019: From Global Ambition to Local Action: A Multi-year Vision for Enhanced Resilience* (Bonn: IGP, 2019).

3. Joanne Linnerooth-Bayer and Stefan Stefan, "Financial Instruments for Disaster Risk Management and Climate Change Adaptation," *Climatic Change* 133, no. 1 (2015): 94, doi:10.1007/s10584-013-1035-6.

4. Paula Jarzabkowski, Konstantinos Chalkias, Daniel Clarke, Ekhosuehi Iyahen, Daniel Stadtmueller, and Astrid Zwick, *Insurance for Climate Adaptation: Opportunities and Limitations* (Rotterdam: Global Commission on Adaptation, 2019), 4.

5. Caribbean Catastrophe Risk Insurance Facility (CCRIF SPC), Company Overview (2020), http://www.ccrif.org.

6. African Risk Capacity, "Government of Senegal to Receive a minimum of US$22m from the African Risk Capacity Insurance Company Limited for Drought," September 5, 2019, https://reliefweb.int/sites/reliefweb.int/files/resources/ARC_PR_Senegal_Payout.pdf.

7. Xiaodong Du, Hongli Feng, and David A. Hennessy, "Rationality of Choices in Subsidized Crop Insurance Markets," *American Journal of Agricultural Economics* 99, no. 3 (2016): 732–756, doi:10.1093/ajae/aaw035.

8. Heather Greatrex, James Hansen, Samantha Garvin, Rahel Diro, Margot Le Guen, Sari Blakely, Kolli Rao, and Dan Osgood, *Scaling Up Index Insurance for Smallholder Farmers: Recent Evidence and Insights* (Copenhagen: CGIAR Research Program on Climate Change, Agriculture and Food Security, 2015).

9. African Risk Capacity, *About* (2019), http://www.arc.int/.

10. Ann Jaworski, "Encouraging Climate Adaptation Through Reform of Federal Crop Insurance Subsidies Notes," *New York University Law Review* 91, no. 6 (2016): 1697.

11. Andrew Crane-Droesch, Elizabeth Marshall, Stephanie Rosch, Anne Riddle, Joseph Cooper, and Steven Wallander, *Climate Change and Agricultural Risk Management Into the 21st Century* (Washington, DC: U.S. Department of Agriculture, 2019).

12. Robin Kundis Craig, "Coastal Adaptation, Government-Subsidized Insurance, and Perverse Incentives to Stay," *Climatic Change* 152, no. 2 (2019): 218, doi:10.1007/s10584-018-2203-5.

13. Maya Dhanjal, "Why Climate Resilience Bonds Can Make a Significant Contribution to Financing Climate Change Adaptation Initiatives," PreventionWeb, United Nations Office for Disaster Risk Reduction, accessed May 2, 2021, https://www.preventionweb.net/go/72119.

14. Sonia Akter, Timothy J. Krupnik, Frederick Rossi, and Fahmida Khanam, "The Influence of Gender and Product Design on Farmers' Preferences for Weather-Indexed Crop Insurance," *Global Environmental Change* 38 (2016): 217–229, doi:10.1016/j.gloenvcha.2016.03.010.

15. InsuResilience Global Partnership, "Mission and Vision" (2020), https://www.insuresilience.org.

16. Linnerooth-Bayer and Stefan, "Financial Instruments," 85–100.

17. Linnerooth-Bayer and Stefan, "Financial Instruments"; Andrea Jonathan Pagano, Maksims Feofilovs, and Francesco Romagnoli, "The Relationship Between Insurance Companies and Natural Disaster Risk Reduction: Overview of the Key Characteristics and Mechanisms Dealing with Climate Change," *Energy Procedia* 147 (2018): 566–572, doi:10.1016/j.egypro.2018.07.072.

18. Linnerooth-Bayer and Stefan, "Financial Instruments," 94.

19. Linnerooth-Bayer and Stefan, "Financial Instruments," 85–100.

20. Kai A. Konrad and Marchel Thum, "The Role of Economic Policy in Climate Change Adaptation," *CESifo Economic Studies* 60, no. 1 (2014): 32–61, doi:10.1093/cesifo/ift003.

21. Konrad and Thum, "The Role of Economic Policy," 46.

7. MIGRATION AND MANAGED RETREAT

1. Nansen Initiative, *Agenda for the Protection of Cross-Border Displaced Persons in the Context of Disasters and Climate Change: Volume 1* (2015), https://disasterdisplacement.org/wp-content/uploads/2014/08/EN _Protection_Agenda_Volume_I_-low_res.pdf.

2. Norman Myers, "Environmental Refugees: An Emergent Security Issue" (Thirteenth Economic Forum, Prague, May 23–27, 2005), https://www.osce.org/files/f/documents/c/3/14851.pdf; Kanta Kumari Rigaud, Alex de Sherbinin, Bryan Jones, Jonas Bergmann, Vivian Clement, Kayly Ober, Jacob Schewe, Susana Adamo, Brent McCusker, Silke Heuser, and Amelia Midgely, *Groundswell: Preparing for Internal Climate Migration* (Washington, DC: World Bank, 2018).

3. Myuki Hino, Christopher B. Field, and Katharine J. Mach, "Managed Retreat as a Response to Natural Hazard Risk," *Nature Climate Change* 7 (2017): 364, doi:10.1038/NCLIMATE3252.

4. U.S. Government Accountability Office (GAO), "Climate Change: A Climate Migration Pilot Program Could Enhance the Nation's Resilience and Reduce Federal Fiscal Exposure" (GAO, Washington, DC, 2020), 1.

5. United Nations, Department of Economic and Social Affairs, Population Division (UN-DESA), *International Migration 2019: Report* (New York: United Nations, 2019), iv.

6. François Gemenne and Julia Blocher, "How Can Migration Serve Adaptation to Climate Change? Challenges to Fleshing Out a Policy Ideal," *Geographic Journal* 183, no. 4 (2017): 336–347, doi:10.1111/geoj.12205.

7. Ine Cottyn, "Livelihood Trajectories in a Context of Repeated Displacement: Empirical Evidence from Rwanda," *Sustainability* 10, no. 10 (2018): 3521, doi:103390/su10103521.

8. A. R. Siders, Miyuki Hino, and Katharine J. Mach, "The Case for Strategic and Managed Climate Retreat," *Science* 365, no. 6455 (2019): 761, doi:10.1126/science.aax8346.

9. Ben Doherty, "'Our Country Will Vanish': Pacific Islanders Bring Desperate Message to Australia," *The Guardian*, May 13, 2017.

10. Silja Klepp and Libertad Chavez-Rodriguez, "Governing Climate Change: The Power of Adaptation Discourses, Policies, and Practices," in *A Critical Approach to Climate Change Adaptation: Discourses, Policies, and Practices*, ed. S. Klepp and L. Chavez-Rodriguez (Oxon, UK: Routledge, 2018), 3.

11. IOM, *World Migration Report*, https://www.un.org/sites/un2.un.org/files/wmr_2020.pdf, 3.

12. Kumari Rigaud et al., *Groundswell*.

13. UN-DESA, *International Migration.*

14. Gemenne and Blocher, "How Can Migration Serve," 343.

15. Siders et al., "The Case for Strategic," 763.

16. Robert Freudenberg, Ellis Calvin, Laura Tolkoff, and Dare Brawley, *Buy-In for Buyouts: The Case for Managed Retreat from Flood Zones* (Cambridge, MA: Lincoln Institute of Land Policy, 2016).

17. Freudenberg et al., *Buy-In for Buyouts,* 7.

18. Robin Kundis Craig, "Coastal Adaptation, Government-Subsidized Insurance, and Perverse Incentives to Stay," *Climatic Change* 152, no. 2 (2019): 222, doi:10.1007/s10584-018-2203-5.

19. Siders et al., "The Case for Strategic," 764.

20. Brian Jones, "Modeling Climate Change-Induced Migration in Central America and Mexico Methodological Report" (2020), https://assets-c3.propublica.org/Climate-Migration-Modeling-Methodology.pdf.

21. Hans Joachim Schellnhuber and Maria A. Martin, "Climate Change, Public Health, Social Peace," in *Health of People, Health of Planet and Our Responsibility*, ed. W. K. Al-Delaimy, V. Ramanathan, and M. Sánchez Sorondo (Cham, Switz.: Springer, 2020), 225–238.

22. Schellnhuber and Martin, "Climate Change."

23. Michael B. Gerrard, "America Is the Worst Polluter in the History of the World. We Should Let Climate Change Refugees Settle Here," *Washington Post*, June 25, 2015.

8. INEQUALITY AND JUSTICE

1. Kris Hartley, "The Epistemics of Policymaking: From Technocracy to Critical Pragmatism in the UN Sustainable Development Goals," *International Review of Public Policy* 2, no. 2 (2020): 233–244, doi:10.4000/irpp.1242.

2. Robert D. Bullard, "The Threat of Environmental Racism," *Natural Resources and Environment* 7, no. 3 (1993): 23–56, https://www.jstor.org/stable/40923229.

3. Yianna Lambrou and Sibyl Nelson, "Farmers in a Changing Climate: Does Gender Matter? Food Security in Andhra Pradesh, India," in *Research, Action, and Policy: Addressing the Gendered Impacts of Climate*

Change, ed. Margaret Alson and Kerri Whittenbury, 189–206 (Rome: Food and Agricultural Organization, 2010).

4. Mizan R. Khan and J. Timmons Roberts, "Towards a Binding Adaptation Regime: Three Levers and Two Instruments," in *Successful Adaptation to Climate Change: Linking Science and Policy in a Rapidly Changing World*, ed. S. C. Moser and M. T. Boykoff (Oxon, UK: Routledge, 2013), 132–148.

5. Hemant R. Ojha, Sharad Ghimire, Adam Pain, Andrea Nightingale, Dil B. Khatri, and Hari Dhungana, "Policy Without Politics: Technocratic Control of Climate Change Adaptation Policy Making in Nepal," *Climate Policy* 16, no. 4 (2016): 423, doi:10.1080/14693062.2014.1003775.

6. Thomas A. Smucker, Ben Wisner, Adolfo Mascarenhas, Pantaleo Munishi, Elizabeth E. Wangui, Gaurav Sinha, Daniel Weiner, Charles Bwenge, and Eric Lovell, "Differentiated Livelihoods, Local Institutions, and the Adaptation Imperative: Assessing Climate Change Adaptation Policy in Tanzania," *Geoforum* 59 (2015): 39–50, doi:10.1016/j.geoforum.2014.11.018.

7. Sophie Webber and Emilia Kennedy, "Climate Change Economies," in *A Critical Approach to Climate Change Adaptation: Discourses, Policies, and Practices*, ed. S. Klepp and L. Chavez-Rodriguez (Oxon, UK: Routledge, 2018), 83.

8. Ojha et al., "Policy Without Politics," 424.

9. Sigrid Nagoda and Andrea J. Nightingale, "Participation and Power in Climate Change Adaptation Policies: Vulnerability in Food Security Programs in Nepal," *World Development* 100 (2017): 85–93, doi:10.1016/j.worlddev.2017.07.022.

10. Edidah L. Ampaire, Laurence Jassogne, Happy Providence, Mariola Acosta, Jennifer Twyman, Leigh Winowiecki, and Piet van Asten, "Institutional Challenges to Climate Change Adaptation: A Case Study on Policy Action Caps in Uganda," *Environmental Science and Policy* 75 (2017): 81–90, doi:10.1016/j.envsci.2017.05.013.

11. Salvador Acquino Centeno, "Ruling Nature and Indigenous Communities," in *A Critical Approach to Climate Change Adaptation: Discourses, Policies, and Practices*, ed. S. Klepp and L. Chavez-Rodriguez (Oxon, UK: Routledge, 2018), 145.

12. Jesse Ribot, "Vulnerability Before Adaptation: Toward Transformative Climate Action," *Global Environmental Change* 21, no. 4 (2011): 1160, doi:10.1016/j.gloenvcha.2011.07.008.

13. Jay Koh, Emilie Mazzacurati, and Stacy Swann, "Bridging the Adaptation Gap: Approaches to Measurement of Physical Climate Risk and Examples of Investment in Climate Adaptation and Resilience" (discussion paper, Global Adaptation and Resilience Investment Working Group, New York, 2016), 13.

14. Johannes F. Linn, "Mobilizing Funds to Combat Climate Change: Lessons from the First Replenishment of the Green Climate Fund," Brookings Institution, February 18, 2020, https://www.brookings.edu/blog/future-development/2020/02/18/mobilizing-funds-to-combat-climate-change-lessons-from-the-first-replenishment-of-the-green-climate-fund/.

15. Mya Sherman, Lea Berrang-Ford, Shuaib Lwasa, James Ford, Didacus B. Namanya, Alejandro Llanos-Cuentes, Michelle Maillet, and Sherilee Harper, "Drawing the Line Between Adaptation and Development: A Systematic Literature Review of Planned Adaptation in Developing Countries," *WIREs Climate Change* 7, no. 5 (2016): 707–726, doi:10.1002/wcc.416.

16. Daniel Morchain, "Rethinking the Framing of Climate Change Adaptation," in *A Critical Approach to Climate Change Adaptation: Discourses, Policies, and Practices*, ed. S. Klepp and L. Chavez-Rodriguez (Oxon, UK: Routledge, 2018), 55–73.

17. Benjamin K. Sovacool, Björn-Ola Linnér, and Richard J. T. Klein, "Climate Change Adaptation and the Least Developed Countries Fund (LDCF): Qualitative Insights from Policy Implementation in the Asia-Pacific," *Climatic Change* 140, no. 2 (2017): 209–226, doi:10.1007/s10584-016-1839-2.

18. Sarah Louise Hemstock, Helene Jacot Des Combes, Leigh-Anne Buliruarua, Kevin Maitova, Ruth Senikula, Roy Smith, and Tess Martin, "Professing the 'Resilience' Sector in the Pacific Islands Region," in *A Critical Approach to Climate Change Adaptation: Discourses, Policies, and Practices*, ed. S. Klepp and L. Chavez-Rodriguez (Oxon, UK: Routledge, 2018), 264.

19. Koh et al., "Bridging the Adaptation Gap," 1–65.

20. Niki Frantzeskaki, Timon McPhearson, Marcus J. Collier, Dave Kendal, Harriet Bulkeley, Adina Dumitru, Claire Walsh, Kate Noble, Ernita van Wyk, Camilo Ordóñez, Cathay Oke, and László Pintér, "Nature-Based Solutions for Urban Climate Change Adaptation: Linking Science,

Policy, and Practice Communities for Evidence-Based Decision-Making,"
BioScience 69, no. 6 (2019): 455–466, doi:10.1093/biosci/biz042.

21. Nathalie Seddon, Alexandre Chausson, Pam Berry, Cécile A. J. Girardin, Alison Smith, and Beth Turner, "Understanding the Value and Limits of Nature-Based Solutions to Climate Change and Other Global Challenges," *Philosophical Transactions of the Royal Society B: Biological Sciences* 375, no. 1794 (2020): 8, doi:10.1098/rstb.2019.0120.

22. Morchain, "Rethinking the Framing," 56.

9. SYNERGIES AND BEST PRACTICES

1. Neil W. Adger, Iain Brown, and Swenja Surminski, "Advances in Risk Assessment for Climate Change Adaptation Policy," *Philosophical Transactions of the Royal Society A: Mathematical, Physical and Engineering Sciences* 376, no. 2121 (2018): 7, doi:10.1098/rsta.2018.0106.

 2. Sara de Wit, "A Clash of Adaptations: How Adaptation to Climate Change Is Translated in Northern Tanzania," in *A Critical Approach to Climate Change Adaptation: Discourses, Policies, and Practices*, ed. S. Klepp and L. Chavez-Rodriguez (Oxon, UK: Routledge, 2018), 43.

 3. Oxfam, "The Vulnerability and Risk Assessment Toolkit," Oxfam, http://vra.oxfam.org.uk/.

 4. Jeff Popke, Scott Curtis, and Douglas W. Gamble, "A Social Justice Framing of Climate Change Discourse and Policy: Adaptation, Resilience and Vulnerability in a Jamaican Agricultural Landscape," *Geoforum* 73 (2016): 72, doi:10.1016/j.geoforum.2014.11.003.

 5. Robert Vautard, Olivier Boucher, Geert Jan van Oldenborgh, Fredeerike Otto, Karsten Haustein, Martha M. Vogel, Sonia I. Seneviratne, Jean-Michel Soubeyroux, Michel Schneider, Agathe Drouin, Arélien Ribes, Frank Kreinkamp, Peter Stott, and Maarten van Aalst, "Human Contribution to the Record-Breaking July 2019 Heat Wave in Western Europe," x, 2019, https://www.worldweatherattribution.org /wp-content/uploads/July2019heatwave.pdf.

 6. Keely Boom, Julie-Anne Richards, and Stephen Leonard, "Climate Justice: The International Momentum Towards Climate Litigation," Heinrich Boell Foundation, 2016, https://www.boell.de/sites/default /files/report-climate-justice-2016.pdf.

BIBLIOGRAPHY

Adger, W. Neil, Iain Brown, and Swenja Surminski. "Advances in Risk Assessment for Climate Change Adaptation Policy." *Philosophical Transactions of the Royal Society A: Mathematical, Physical and Engineering Sciences* 376, no. 2121 (2018): 1–13. doi:10.1098/rsta.2018.0106.

Akter, Sonia, Timothy J. Krupnik, Frederick Rossi, and Fahmida Khanam. "The Influence of Gender and Product Design on Farmers' Preferences for Weather-Indexed Crop Insurance." *Global Environmental Change* 38 (2016): 217–229. doi:10.1016/j.gloenvcha.2016.03.010.

Ampaire, Edidah L., Laurence Jassogne, Happy Providence, Mariola Acosta, Jennifer Twyman, Leigh Winowiecki, and Piet van Asten. "Institutional Challenges to Climate Change Adaptation: A Case Study on Policy Action Gaps in Uganda." *Environmental Science & Policy* 75 (2017): 81–90. doi:10.1016/j.envsci.2017.05.013.

Araos, Malcolm, Lea Berrang-Ford, James D. Ford, Stephanie E. Austin, Robert Biesbroek, and Alexandra Lesnikowski. "Climate Change Adaptation Planning in Large Cities: A Systematic Global Assessment." *Environmental Science & Policy* 66 (2016): 375–382. doi:10.1016/j.envsci.2016.06.009.

Archibald, Carla L., and Nathalie Butt. "Using Google Search Data to Inform Global Climate Change Adaptation Policy." *Climate Change* 150, no. 3 (2018): 447–456. doi:10.1007/s10584-018-2289-9.

Artur, Luis, and Dorothea Hilhorst. "Everyday Realities of Climate Change Adaptation in Mozambique." *Global Environmental Change* 22, no. 2 (2012): 529–536. doi:10.1016/j.gloenvcha.2011.11.013.

Baba, Kenshi, and Mitsuru Tanaka. "Attitudes of Farmers and Rural Area Residents Toward Climate Change Adaptation Measures: Their Preferences and Determinants of Their Attitudes." *Climate* 7, no. 5 (2019): 1–11. doi:10.3390/cli7050071.

Balaban, Osman, and Meltem Şenol Balaban. "Adaptation to Climate Change: Barriers in the Turkish Local Context." *TeMA: Journal of Land Use, Mobility and Environment*, no. 8 special issue ECCA (2015): 7–22. doi:10.6092/1970-9870/3650.

Bangalore, Mukund Ram. *Shock Waves: Managing the Impacts of Climate Change on Poverty*. Washington, D.C.: World Bank Group, 2016.

Bankoff, Greg. "Time Is of the Essence: Disasters, Vulnerability, and History." *International Journal of Mass Emergencies and Disasters* 22, no. 3 (2004): 23–42.

Batisani, Nnyaladzi, and Brent Yarnel. "Rainfall Variability and Trends in Semi-Arid Botswana: Implications for Climate Change Adaptation Policy." *Applied Geography* 30, no. 4 (2010): 483–489. doi:10.1016/j.apgeog .2009.10.007.

Bécault, Emilie, Moritz Koenig, and Axel Marx. *Getting Ready for Climate Finance: The Case of Rwanda*. Leuven: Belgian Policy Research Group on Financing for Development, 2016.

Bele, Mekou Youssoufa, Olufunso Somorin, Denis Jean Sonwa, Johnson Ndi Nkem, and Bruno Locatelli. "Forests and Climate Change Adaptation Policies in Cameroon." *Mitigation and Adaptation Strategies for Global Change* 16, no. 3 (2011): 369–385. doi:10.1007/s11027-010-9264-8.

Béné, Christophe, Derek Headey, Lawrence Haddad, and Klaus von Grebmer. "Is Resilience a Useful Concept in the Context of Food Security and Nutrition Programmes? Some Conceptual and Practical Considerations." *Food Security* 8, no. 1 (2016): 123–138. doi:10.1007 /s12571-015-0526-x.

Berkhout, Frans, Laurens M. Bouwer, Joanne Bayer, Maha Bouzid, Mar Cabeza, Susanne Hanger, Andries Hof, Paul Hunter, Laura Meller, Anthony Patt, Benjamin Pfluger, Tim Rayner, Kristin Reichardt, and Astrid van Teeffelen. "European Policy Responses to Climate Change: Progress on Mainstreaming Emissions Reduction and Adaptation." *Regional Environmental Change* 15, no. 6 (2015): 949–959. doi:10.1007 /s10113-015-0801-6.

Bhullar, Lovleen. "Climate Change Adaptation and Water Policy: Lessons from Singapore." *Sustainable Development* 21, no. 3 (2013): 152–159. doi:10.1002/sd.1546.

Biermann, Frank, and Ingrid Boas. "Protecting Climate Refugees: The Case for a Global Protocol." *Environment* 50 (2008): 8–17. doi:10.3200/ENVT.50.6.8-17.

Biesbroek, Robert, Lea Berrang-Ford, James D. Ford, Andrew Tanabe, Stephanie E. Austin, and Alexandra Lesnikowski. "Data, Concepts and Methods for Large-*n* Comparative Climate Change Adaptation Policy Research: A Systematic Literature Review." *WIREs Climate Change* 9, no. 6 (2018): 1–15. doi:10.1002/wcc.548.

Biesbroek, Robert, Alexandra Lesnikowski, James D. Ford, Lea Berrang-Ford, and Martinus Vink. "Do Administrative Traditions Matter for Climate Change Adaptation Policy? A Comparative Analysis of 32 High-Income Countries." *Review of Policy Research* 35, no. 6 (2018): 881–906. doi:10.1111/ropr.12309.

Biesbroek, Robbert, and Jereon J. L. Candel. "Mechanisms for Policy (Dis) integration: Explaining Food Policy and Climate Change Adaptation Policy in the Netherlands." *Policy Sciences; Dordrecht* 53, no. 1 (2020): 61–84. doi:10.1007/s11077-019-09354-2.

Bonzanigo, Laura, Dragana Bojovic, Alexandros Maziotis, and Carlo Giupponi. "Agricultural Policy Informed by Farmers' Adaptation Experience to Climate Change in Veneto, Italy." *Regional Environmental Change* 16, no. 1 (2016): 245–258. doi:10.1007/s10113-014-0750-5.

Bosnjakovic, Branko, and Iva Mrsa Haber. "Climate Changes and Adaptation Policies in the Baltic and the Adriatic Regions." *UTMS Journal of Economics* 6, no. 1 (2015): 21–39.

Chu, Eric K. "The Governance of Climate Change Adaptation through Urban Policy Experiments." *Environmental Policy and Governance* 26, no. 6 (2016): 439–451. doi:10.1002/eet.1727.

Ciplet, David, J. Timmons Roberts, and Mizan Khan. "The Politics of International Climate Adaptation Funding: Justice and Divisions in the Greenhouse." *Global Environmental Politics* 13, no. 1 (2013): 49–68. doi: 10.1162/GLEP_a_00153.

Clar, Christoph, Andrea Prutsch, and Reinhard Steurer. "Barriers and Guidelines for Public Policies on Climate Change Adaptation: A Missed

Opportunity of Scientific Knowledge-Brokerage." *Natural Resources Forum* 37, no. 1 (2013): 1–18. doi:10.1111/1477-8947.12013.

Clar, Christoph, and Reinhard Steurer. "Why Popular Support Tools on Climate Change Adaptation Have Difficulties in Reaching Local Policy-Makers: Qualitative Insights from the UK and Germany." *Environmental Policy and Governance* 28, no. 3 (2018): 172–182. doi:10.1002/eet.1802.

Clay, Nathan, and Brian King. "Smallholders' Uneven Capacities to Adapt to Climate Change Amid Africa's 'Green Revolution': Case Study of Rwanda's Crop Intensification Program." *World Development* 116 (2019): 1–14. doi:10.1016/j.worlddev.2018.11.022.

Coffey, Peter, and Robert Riley. *Reform of the International Institutions.* Northampton, MA: Edward Elgar, 2006.

Craig, Robin Kundis. "Coastal Adaptation, Government-Subsidized Insurance, and Perverse Incentives to Stay." *Climatic Change* 152, no. 2 (2019): 215–226. doi:10.1007/s10584-018-2203-5.

D'Aquino, Patrick, and Alassane Bah. "Land Policies for Climate Change Adaptation in West Africa: A Multilevel Companion Modeling Approach." *Simulation & Gaming* 44, no. 2–3 (2013): 391–408. doi:10.1177/1046878112452689.

De Scherbinin, Alex, Andrew Schiller, and Alex Pulsipher. "The Vulnerability of Global Cities to Climate Hazards." *Environment and Urbanization* 19, no. 1 (2007): 39–64. doi:10.1177/0956247807076725.

Derkzen, Marthe L., Astrid J. A. van Teefelen, and Peter H. Verburg. "Green Infrastructure for Urban Climate Adaptation: How Do Residents' Views on Climate Impacts and Green Infrastructure Shape Adaptation Preferences?" *Landscape and Urban Planning* 157 (2017): 106–130. doi:10.1016/j.landurbplan.2016.05.027.

Donaldson, Dave, Arnaud Costinot, and Cory B. Smith. *Evolving Comparative Advantage and the Impact of Climate Change in Agricultural Markets: Evidence from 1.7 Million Fields around the World.* Cambridge, MA: National Bureau of Economic Research, 2014.

Dupuis, Johann, and Robbert Biesbroek. "Comparing Apples and Oranges: The Dependent Variable Problem in Comparing and Evaluating Climate Change Adaptation Policies." *Global Environmental Change* 23, no. 6 (2013): 1476–1487. doi:10.1016/j.gloenvcha.2013.07.022.

Dupuis, Johann, and Peter Knoepfel. "The Adaptation Policy Paradox: The Implementation Deficit of Policies Framed as Climate Change

Adaptation." *Ecology and Society* 18, no. 4 (2013): 1–16. doi:10.5751/ES-05965-180431.

Elliott, James R., and Jeremy Pais. "Race, Class, and Hurricane Katrina: Social Differences in Human Responses to Disaster." *Social Science Research* 35, no. 2 (2006): 295–321. doi:10.1016/j.ssresearch.2006.02.003.

England, Matthew I., Andrew J. Dougill, Lindsay C. Stringer, Katherine E. Vincent, Joanna Pardoe, Felix K. Kalaba, David D. Mkwambisi, Emilinah Namaganda, and Stavros Afionis. "Climate Change Adaptation and Cross-Sectoral Policy Coherence in Southern Africa." *Regional Environmental Change* 18, no. 7 (2018): 2059–2071. doi:10.1007/s10113-018-1283-0.

Filho, Walter Leal, Abdul-Lateef Balogun, Desalegn Yayeh Ayal, E. Matthew Bethurem, Miriam Murambadoro, Julia Mambo, Habitamu Taddese, Gebrekidan Worku Tefera, Gustavo J. Nagy, Hubert Fudjumdjum, and Paschal Mugabe. "Strengthening Climate Change Adaptation Capacity in Africa: Case Studies from Six Major African Cities and Policy Implications." *Environmental Science & Policy* 86 (2018): 29–37. doi:10.1016/j.envsci.2018.05.004.

Forino, Giuseppe, Jason von Meding, Graham Brewer, and Thayaparan Gajendran. "Disaster Risk Reduction and Climate Change Adaptation Policy in Australia." *Procedia Economics and Finance* 18 (2014): 473–482. doi:10.1016/S2212-5671(14)00965-4.

Frantzeskaki, Niki, Timon McPhearson, Marcus J. Collier, Dave Kendal, Harriet Bulkeley, Adina Dumitru, Claire Walsh, Kate Noble, Ernita van Wyk, Camilo Ordóñez, Cathay Oke, and László Pintér. "Nature-Based Solutions for Urban Climate Change Adaptation: Linking Science, Policy, and Practice Communities for Evidence-Based Decision-Making." *BioScience* 69, no. 6 (2019): 455–466. doi:10.1093/biosci/biz042.

Freudenberg, Robert, Ellis Calvin, Laura Tolkoff, and Dare Brawley. *Buy-in for Buyouts: The Case for Managed Retreat from Flood Zones.* Cambridge, MA: Lincoln Institute of Land Policy, 2016.

Funder, Mikkel, and Carol Emma Mweemba. "Interface Bureaucrats and the Everyday Remaking of Climate Interventions: Evidence from Climate Change Adaptation in Zambia." *Global Environmental Change* 55 (2019): 130–138. doi:10.1016/j.gloenvcha.2019.02.007.

Gemenne, François, and Julia Blocher. "How Can Migration Serve Adaptation to Climate Change? Challenges to Fleshing Out a Policy Ideal." *Geographic Journal* 183, no. 4 (2017): 336–347. doi:10.1111/geoj.12205.

Greatrex, Heather, James Hansen, Samantha Garvin, Rahel Diro, Margot Le Guen, Sari Blakely, Kolli Rao, and Dan Osgood. *Scaling Up Index Insurance for Smallholder Farmers: Recent Evidence and Insights.* Copenhagen: CGIAR Research Program on Climate Change, Agriculture and Food Security, 2015.

Hahn, Micha B., Anne M. Riederer, and Stanley O. Foster. "The Livelihood Vulnerability Index: A Pragmatic Approach to Assessing Risks from Climate Variability and Change: A Case Study in Mozambique." *Global Environmental Change* 19, no. 1 (2009): 74–88. doi:10.1016/j.gloenvcha.2008.11.002.

Harvey, Chelsea. "Scientists Can Now Blame Individual Natural Disasters on Climate Change." *ClimateWire*, January 2, 2018.

Holmes, Rebecca, Kay Sharp, Nathalie Both, and Roberte Isimbi. *Qualitative Evaluation of the Vision 2020 Umurenge Programme Safety Net.* Oxford: High-Quality Technical Assistance for Results, 2018.

Howarth, C., S. Morse-Jones, K. Brooks, and A. P. Kythreotis. "Co-Producing UK Climate Change Adaptation Policy: An Analysis of the 2012 and 2017 UK Climate Change Risk Assessments." *Environmental Science & Policy* 89 (2018): 412–420. doi:10.1016/j.envsci.2018.09.010.

Hussain, Abid, Nand Kishor Agrawal, and Iris Leikanger. "Action for Adaptation: Bringing Climate Change Science to Policy Makers—A Synthesis Report of a Conference Held in Islamabad on 23–25 July 2015." *Food Security* 8, no. 1 (2016): 285–289. doi:10.1007/s12571-015-0529-7.

Ingold, Karin. "How to Create and Preserve Social Capital in Climate Adaptation Policies: A Network Approach." *Ecological Economics* 131 (2017): 414–424. doi:10.1016/j.ecolecon.2016.08.033.

Jäger, J., M. D. A. Rounsevell, P. A. Harrison, I. Omann, R. Dunford, M. Kammerlander, and G. Pataki. "Assessing Policy Robustness of Climate Change Adaptation Measures across Sectors and Scenarios." *Climatic Change* 128, no. 3 (2015): 395–407. doi:10.1007/s10584-014-1240-y.

James, Rachel, Friederike Otto, Hannah Parker, Emily Boyd, Rosalind Cornforth, Daniel Mitchell, and Myles Allen. "Characterizing Loss and Damage from Climate Change." *Nature Climate Change* 4, no. 11 (2014): 938–939. doi:10.1038/nclimate2411.

Jarzabkowski, Paula, Konstantinos Chalkias, Daniel Clarke, Ekhosuehi Iyahen, Daniel Stadtmueller, and Astrid Zwick. *Insurance for Climate*

Adaptation: Opportunities and Limitations. Rotterdam: Global Commission on Adaptation, 2019.

Jaworski, Ann. "Encouraging Climate Adaptation through Reform of Federal Crop Insurance Subsidies Notes." *New York University Law Review* 91, no. 6 (2016): 1684–1718.

Kahsey, Goytom Abraha, and Lars Gårn Hansen. "The Effect of Climate Change and Adaptation Policy on Agricultural Production in Eastern Africa." *Ecological Economics* 121 (2016): 54–64. doi:10.1016/j.ecolecon.2015.11.016.

Kalabamu, Faustin Tirwirukwa. "Land Tenure Reforms and Persistence of Land Conflicts in Sub-Saharan Africa: The Case of Botswana." *Land Use Policy* 81 (2019): 337–345. doi:10.1016/j.landusepol.2018.11.002.

Kelman, Ilan. "Lost for Words amongst Disaster Risk Science Vocabulary?" *International Journal of Disaster Risk Science* 9, no. 3 (2018): 281–291. doi:10.1007/s13753-018-0188-3.

Keskitalo, E. Carina H., M. Legay, M. Marchetti, S. Nocentini, and P. Spathelf. "The Role of Forestry in National Climate Change Adaptation Policy: Cases from Sweden, Germany, France and Italy." *International Forestry Review* 17, no. 1 (2015): 30–42. doi:10.2307/24310650.

Keskitalo, E. Carina H., Gregor Vulturius, and Peter Scholten. "Adaptation to Climate Change in the Insurance Sector: Examples from the UK, Germany, and the Netherlands." *Natural Hazards* 71, no. 1 (2014): 315–334. doi:10.1007/s11069-013-0912-7.

Kielkowska, Julianna, Katarzyna Tokarczyk-Dorociak, Jan K. Kazak, Szymon Szewranski, and Joost van Hoof. "Urban Adaptation to Climate Change Plans and Policies." *Journal of Ecological Engineering* 19, no. 2 (2018): 50–62. doi:10.12911/22998993/81658.

Kim, Jin-Oh, and Joo-Hwan Suh. "A Review of Climate Change Adaptation Policies Applied to Landscape Planning and Design in Korea." *Landscape and Ecological Engineering* 12, no. 1 (2016): 171–177. doi:10.1007/s11355-014-0261-z.

Konrad, Kai A., and Marchel Thum. "The Role of Economic Policy in Climate Change Adaptation." *CESifo Economic Studies* 60, no. 1 (2014): 32–61. doi:10.1093/cesifo/ifto03.

Kotecký, Vojtěch. "Contribution of Afforestation Subsidies Policy to Climate Change Adaptation in the Czech Republic." *Land Use Policy* 47 (2015): 112–120. doi:10.1016/j.landusepol.2015.03.014.

Larsen, Rasmus Kløcker, Åsa Gerger Swartling, Neil Powell, Brad May, Ryan Plummer, Louise Simonsson, and Maria Osbeck. "A Framework for Facilitating Dialogue between Policy Planners and Local Climate Change Adaptation Professionals: Cases from Sweden, Canada, and Indonesia." *Environmental Science & Policy* 23 (2012): 12–23. doi:10.1016/j.envsci.2012.06.014.

Linnerooth-Bayer, Joanne, and Stefan Stefan. "Financial Instruments for Disaster Risk Management and Climate Change Adaptation." *Climatic Change* 133, no. 1 (2015): 85–100. doi:10.1007/s10584-013-1035-6.

Locatelli, Bruno, Charlotte Pavageau, Emilia Pramova, and Monica Di Gregorio. "Integrating Climate Change Mitigation and Adaptation in Agriculture and Forestry: Opportunities and Trade-Offs." *WIREs Climate Change* 6, no. 6 (2015): 585–598. doi:10.1002/wcc.357.

Lungarska, Anna, and Raja Chakir. "Climate-Induced Land Use Change in France: Impacts of Agricultural Adaptation and Climate Change Mitigation." *Ecological Economics* 147 (2018): 134–154. doi:10.1016/j.ecolecon.2017.12.030.

Lwasa, Shuaib. "A Systematic Review of Research on Climate Change Adaptation Policy and Practice in Africa and South Asia Deltas." *Regional Environmental Change* 15, no. 5 (2015): 815–824. doi:10.1007/s10113-014-0715-8.

Manocha, Nishtha, and Vladan Babovic. "Planning Flood Risk Infrastructure Development Under Climate Change Uncertainty." *Procedia Engineering* 154 (2016): 1406–1413. doi:10.1016/j.proeng.2016.07.511

Martins, Rafael D'Almeida. "International Research Institute for Climate and Society." In *Encyclopedia of Global Warming & Climate Change*, ed. S. George Philander, 971–791. Thousand Oaks, CA: Sage, 2012.

Massey, Eric, Robbert Biesbroek, Dave Huitema, and Andy Jordan. "Climate Policy Innovation: The Adoption and Diffusion of Adaptation Policies Across Europe." *Global Environmental Change* 29 (2014): 434–443. doi:10.1016/j.gloenvcha.2014.09.002.

Masuda, Yuta J., Brianna Castro, Ike Aggraeni, Nicholas H. Wolff, Kristie Ebi, Teevrat Garg, Edward T. Game, Jennifer Krenz, and June Spector. "How Are Healthy, Working Populations Affected by Increasing Temperatures in the Tropics? Implications for Climate Change Adaptation Policies." *Global Environmental Change* 56 (2019): 29–40. doi:10.1016/j.gloenvcha.2019.03.005.

Mees, Heleen L. P., Justin Dijk, Daan van Soest, Peter P. J. Driessen, Marleen H. F. M. W. van Rijswick, and Hens Runhaar. "A Method for the Deliberate and Deliberative Selection of Policy Instrument Mixes for Climate Change Adaptation." *Ecology and Society* 19, no. 2 (2014): 1–16. doi:10.5751 /ES-06639-190258.

Mercer, Jessica, Ilan Kelman, Francisco de Rosario, Abilio de Deus de Jesus Lima, Augusto de Silva, Anna-Maija Beloff, and Alex McClean. "Nation-Building Policies in Timor-Leste: Disaster Risk Reduction, Including Climate Change Adaptation." *Disasters* 38, no. 4 (2014): 690–718. doi:10.1111 /disa.12082.

Mikova, Kseniia, Encok Makupa, and John Kayumba. "Effect of Climate Change on Crop Production in Rwanda." *Earth Sciences* 4, no. 3 (2015): 120–128. doi:10.11648/j.earth.20150403.15.

Millard, Bill. "Symposium: Extreme Heat: Hot Cities—Adapting to a Hotter World." *The Design for Risk and Reconstruction Committee of the American Institute of Architects' New York Chapter*, November 12, 2015. https://www.yumpu.com/xx/document/read/55505322/extreme-heat-hot -cities-adapting-to-a-hotter-world.

Musah-Surugu, Issah Justice, Albert Ahenkan, and Justice Nyigmah Bawole. "Too Weak to Lead: Motivation, Agenda Setting, and Constraints of Local Government to Implement Decentralized Climate Change Adaptation Policy in Ghana." *Environment, Development and Sustainability; Dordrecht* 21, no. 2 (2017): 587–607. doi:10.1007/s10668-017-0049-z.

Musah-Surugu, Issah Justice, Albert Ahenkan, Justice Nyigmah Bawole, and Samuel Antwi Darkwah. "Migrants' Remittances: A Complementary Source of Financing Adaptation to Climate Change at the Local Level in Ghana." *International Journal of Climate Change Strategies and Management* 10, no. 1 (2018): 178–196. doi:10.1108/IJCCSM-03-2017-0054.

Nada, Lazarević-Bajec. "Integrating Climate Change Adaptation Policies in Spatial Development Planning in Serbia: A Challenging Task Ahead." *Spatium* 24 (2011): 1–8. doi:10.2298/SPAT1124001L.

Naess, Lars Otto, Peter Newell, Andrew Newsham, Jon Phillips, Julian Quan, and Thomas Tanner. "Climate Policy Meets National Development Contexts: Insights from Kenya and Mozambique." *Global Environmental Change* 35 (2015) 534–544. doi:10.1016/j.gloenvcha.2015.08.015.

Nagoda, Sigrid. "New Discourses but Same Old Development Approaches? Climate Change Adaptation Policies, Chronic Food Insecurity and

Development Interventions in Northwestern Nepal." *Global Environmental Change* 35 (2015): 570–579. doi:10.1016/j.gloenvcha.2015.08.014.

Nagoda, Sigrid, and Andrea J. Nightingale. "Participation and Power in Climate Change Adaptation Policies: Vulnerability in Food Security Programs in Nepal." *World Development* 100 (2017): 85–93. doi:10.1016/j.worlddev.2017.07.022.

Ng, Kiat, Inês Campos, and Gil Penha-Lopes, eds. *BASE Adaptation Inspiration Book: 23 European Cases of Climate Change Adaptation to Inspire European Decision-Makers, Practioners, and Citizens.* Lisbon: Faculty of Sciences, University of Lisbon, 2016.

Nguyen, Thu Thi, Jamie Pittock, and Bich Huong Nguyen. "Integration of Ecosystem-Based Adaptation to Climate Change Policies in Viet Nam." *Climatic Change; Dordrecht* 142, no. 1–2 (2017): 97–111. doi:10.1007/s10584-017-1936-x.

Nyasimi, Mary, Maren A. O. Radeny, and James Hansen. *Review of Climate Service Needs and Opportunities in Rwanda.* Copenhagen: CGIAR Research Program on Climate Change, Agriculture and Food Security, 2016.

Oija, Hemant R., Sharad Ghimire, Adam Pain, Andrea Nightingale, Dil B. Khatri, and Hari Dhungana. "Policy without Politics: Technocratic Control of Climate Change Adaptation Policy Making in Nepal." *Climate Policy* 16, no. 4 (2016): 415–433. doi:10.1080/14693062.2014.1003775.

Osbahr, Henny, Chasca Twyman, W. Neil Adger, and David S. G. Thomas. "Effective Livelihood Adaptation to Climate Change Disturbance: Scale Dimensions of Practice in Mozambique." *Geoforum* 39, no. 6 (2008): 1951–1964. doi:10.1016/j.geoforum.2008.07.010.

Pagano, Andrea Jonathan, Maksims Feofilovs, and Francesco Romagnoli. "The Relationship Between Insurance Companies and Natural Disaster Risk Reduction: Overview of the Key Characteristics and Mechanisms Dealing with Climate Change." *Energy Procedia* 147 (2018): 566–572. doi:10.1016/j.egypro.2018.07.072.

Papathoma-Köhle, Marie, Catrin Promper, and Thomas Glade. "A Common Methodology for Risk Assessment and Mapping of Climate Change Related Hazards—Implications for Climate Change Adaptation Policies." *Climate* 4, no. 1 (2016): 1–23. doi:10.3390/cli4010008.

Picketts, Ian M. "Practitioners, Priorities, Plans, and Policies: Assessing Climate Change Adaptation Actions in a Canadian Community." *Sustainability Science* 10, no. 3 (2015): 503–513. doi:10.1007/s11625-014-0271-7.

Pietrapertosa, Filomena, Valeriy Khokhlov, Monica Salvia, and Carmelina Cosmi. "Climate Change Adaptation Policies and Plans: A Survey in 11 South East European Countries." *Renewable and Sustainable Energy Reviews* 81 (2018): 3041–3050. doi:10.1016/j.rser.2017.06.116.

Popke, Jeff, Scott Curtis, and Douglas W. Gamble. "A Social Justice Framing of Climate Change Discourse and Policy: Adaptation, Resilience and Vulnerability in a Jamaican Agricultural Landscape." *Geoforum* 73 (2016): 70–80. doi:10.1016/j.geoforum.2014.11.003.

Prutsch, Andrea, Torsten Grothmann, Sabine McCallum, Inke Schauser, and Rob Swart. *Climate Change Adaptation Manual: Lessons Learned from European and Other Industrialized Countries.* Abingdon: Routledge, 2014.

Quinn, Colin F., Jennifer F. Howard, Chen Chen, Joyce E. Coffee, Carlos E. Quintela, Britt A. Parker, and Joel B. Smith. "Adaptation and Poverty Reduction in Mozambique: An Opportunity for Developing Countries to Lead." *Climate Policy* 18, no. 2 (2018): 146–150. doi:10.1080/14693062. 2016.1258631.

Ranabat, Sunita, Rucha Ghate, Laxmi Dutt Bhatta, Nand Kishor Agrawal, and Sunil Tankha. "Policy Coherence and Interplay between Climate Change Adaptation Policies and the Forestry Sector in Nepal." *Environmental Management; New York* 61, no. 6 (2018): 968–980. doi:10.1007 /s00267-018-1027-4.

Rayner, Jeremy, Kathleen McNutt, and Adam Wellstead. "Dispersed Capacity and Weak Coordination: The Challenge of Climate Change Adaptation in Canada's Forest Policy Sector." *Review of Policy Research* 30, no. 1 (2013): 66–90. doi:10.1111/ropr.12003.

Reguero, Borja G., Michael W. Beck, David N. Bresch, Juliano Calil, and Imen Meliane. "Comparing the Cost Effectiveness of Nature-Based and Coastal Adaptation: A Case Study from the Gulf Coast of the United States." *PLoS ONE* 13, no. 4 (2018): 1–24. doi:10.1371/journal .pone.0192132.

Republic of Rwanda. *Green Growth and Climate Resilience: National Strategy for Climate Change and Low Carbon Development.* Kigali, Rwanda, 2011.

Rezende, Camila Linhares, Joana Stingel Fraga, Juliana Cabral Sessa, Gustavo Vinagre Pinto de Souza, Eduardo Delgado Assad, and Fabio Rubio Scarano. "Land Use Policy as a Driver for Climate Change Adaptation: A Case in the Domain of the Brazilian Atlantic Forest." *Land Use Policy* 72 (2018): 563–569. doi:10.1016/j.landusepol.2018.01.027.

Ritchie, Hannah, and David Roser. "Natural Disasters." *Our World in Data*, 2014, https://ourworldindata.org/natural-disasters.

Rondhi, Mohammed, Ahmad Fatikhul Khasen, Yasuhiro Mori, and Takumi Kondo. "Assessing the Role of the Perceived Impact of Climate Change on National Adaptation Policy: The Case of Rice Farming in Indonesia." *Land* 8, no. 5 (2019): 1–21. doi:10.3390/land8050081.

Schmidt, Nicole M., Na'ama Teschner, and Maya Negev. "Scientific Advice and Administrative Traditions: The Role of Chief Scientists in Climate Change Adaptation." *Review of Policy Research* 35, no. 6 (2018): 859–880. doi:10.1111/ropr.12295.

Seddon, Nathalie, Alexandre Chausson, Pam Berry, Cécile A. J. Girardin, Alison Smith, and Beth Turner. "Understanding the Value and Limits of Nature-Based Solutions to Climate Change and Other Global Challenges." *Philosophical Transactions of the Royal Society B: Biological Sciences* 375, no. 1794 (2020): 1–12. doi:10.1098/rstb.2019.0120.

Sherman, Mya, Lea Berrang-Ford, Shuaib Lwasa, James Ford, Didacus B. Namanya, Alejandro Llanos-Cuentes, Michelle Maillet, and Sherilee Harper. "Drawing the Line between Adaptation and Development: A Systematic Literature Review of Planned Adaptation in Developing Countries." *WIREs Climate Change* 7, no. 5 (2016): 707–726. doi:10.1002/wcc.416.

Sietz, Diana, Maria Boschütz, and Richard J. T. Klein. "Mainstreaming Climate Adaptation into Development Assistance: Rationale, Institutional Barriers, and Opportunities in Mozambique." *Environmental Science & Policy* 14, no. 4 (2011): 493–502. doi:10.1016/j.envsci.2011.01.001.

Singh, Ajay S., Adam Zwickle, Jeremy T. Bruskotter, and Robyn Wilson. "The Perceived Psychological Distance of Climate Change Impacts and Its Influence on Support for Adaptation Policy." *Environmental Science & Policy* 73 (2017): 93–99. doi:10.1016/j.envsci.2017.04.011.

Smucker, Thomas A., Ben Wisner, Adolfo Mascarenhas, Pantaleo Munishi, Elizabeth E. Wangui, Gaurav Sinha, Daniel Weiner, Charles Bwenge, and Eric Lovell. "Differentiated Livelihoods, Local Institutions, and the Adaptation Imperative: Assessing Climate Change Adaptation Policy in Tanzania." *Geoforum* 59 (2015): 39–50. doi:10.1016/j.geoforum.2014.11.018.

Sörensen, J., and T. Emilsson. "Evaluating Flood Risk Reduction by Urban Blue-Green Infrastructure Using Insurance Data." *Journal of Water*

Resources Planning and Management 145, no. 2 (2019). doi: 10.1061/(ASCE) WR.1943-5452.0001037.

Sosa-Rodriguez, Fabiola S. "From Federal to City Mitigation and Adaptation: Climate Change Policy in Mexico City." *Mitigation and Adaptation Strategies for Global Change* 19, no. 7 (2014): 969–996. doi:10.1007 /s11027-013-9455-1.

Sovacool, Benjamin K., Björn-Ola Linnér, and Richard J. T. Klein. "Climate Change Adaptation and the Least Developed Countries Fund (LDCF): Qualitative Insights from Policy Implementation in the Asia-Pacific." *Climatic Change* 140, no. 2 (2017): 209–226. doi:10.1007/s10584-016-1839-2.

Stadelmann, Martin, Axel Michaelowa, Sonja Butzengeiger-Geyer, and Michel Köhler. "Universal Metrics to Compare the Effectiveness of Climate Change Adaptation Projects." In *Handbook of Climate Change Adaptation*, ed. W. Leal Filho, 152–174 (Berlin: Springer-Verlag, 2015). doi:10.1007/978-3-642-38670-1_99.

Stringer, Lindsay C., Andrew J. Dougill, Jen C. Dyer, Katharine Vincent, Florian Fritzsche, Julia Leventon, Mario Paulo Falcão, Pacal Manyakaidze, Stephen Syampungani, Philip Powell, and Gabriel Kalaba. "Advancing Climate Compatible Development: Lessons from Southern Africa." *Regional Environmental Change* 14, no. 2 (2014): 713–725. doi:10.1007 /s10113-013-0533-4.

Sussams, L. W., W. R. Sheate, and R. P. Eales. "Green Infrastructure as a Climate Change Adaptation Policy Intervention: Muddying the Waters or Clearing a Path to a More Secure Future?" *Journal of Environmental Management* 147 (2015): 184–193. doi:10.1016/j.jenvman.2014.09.003.

Terminski, Bogumil. *Development-Induced Displacement and Resettlement: Causes, Consequences, and Socio-Legal Context.* New York: Columbia University Press, 2014.

Thaler, Thomas, Sven Fuchs, Sally Priest, and Neelke Doorn. "Social Justice in the Context of Adaptation to Climate Change: Reflecting on Different Policy Approaches to Distribute and Allocate Flood Risk Management." *Regional Environmental Change* 18, no. 2 (2018): 305–309. doi:10.1007 /s10113-017-1272-8.

Timberlake, Thomas J., and Courtney A. Schultz. "Policy, Practice, and Partnerships for Climate Change Adaptation on US National Forests." *Climatic Change* 144, no. 2 (2017): 257–269. doi:10.1007/s10584-017-2031-z.

United Nations Development Programme. *Increasing Climate-Resilience in Rwanda through EWS, Disaster Preparedness, and Integrated Watershed Management*. 2012. https://www.adaptation-undp.org/sites/default/files /downloads/alm_case_study_rwanda_dec2012.pdf.

Urwin, Kate, and Andrew Jordan. "Does Public Policy Support or Undermine Climate Change Adaptation? Exploring Policy Interplay across Different Scales of Governance." *Global Environmental Change* 18, no. 1 (2008): 180–191. doi:10.1016/j.gloenvcha.2007.08.002.

Van Aalst, Maarten K., Terry Cannon, and Ian Burton. "Community-Level Adaptation to Climate Change: The Potential Role of Participatory Community Risk Assessment." *Global Environmental Change* 18, no. 1 (2008); 165–179. doi:10.1016/j.gloenvcha.2007.06.002.

Vij, Sumit, Eddy Moors, Bashir Ahmad, Md. Arfanuzzaman, Suruchi Bhadwal, Robbert Biesbroek, Giovanna Gioli, Annemarie Groot, Dwijen Mallick, Bimal Regmi, Basharat Ahmed Saeed, Sultan Ishaq, Bhuwan Thapa, Saskia E. Werners, and Phillipus Wester. "Climate Adaptation Approaches and Key Policy Characteristics: Cases from South Asia." *Environmental Science & Policy* 78 (2017): 58–65. doi:10.1016 /j.envsci.2017.09.007.

Wamsler, C., B. Wickenberg, H. Hanson, J. Alkan-Olsson, S. Stålhammar, H. Björn, H. Falck, D. Gerell, T. Oskarsson, E. Simonsson, F. Torffvit, and F. Zelmerlow. "Environmental and Climate Policy Integration: Targeted Strategies for Overcoming Barriers to Nature-Based Solutions and Climate Change Adaptation." *Journal of Cleaner Production* 247 (2020): 1–10. doi:10.1016/j.jclepro.2019.119154.

Warner, K., P. van de Logt, M. Brouwer, A. J. van Bodegom, B. Satijn, F. M. Galema, and G. L. Buit. *Climate Change Profile: Rwanda*. Utrecht: Netherlands Commission for Environmental Assessment, Dutch Sustainability Unit, 2015.

Webber, Sophie. "Performative Vulnerability: Climate Change Adaptation Policies and Financing in Kiribati." *Environment and Planning A: Economy and Space* 45, no. 11 (2013): 2717–2733. doi:10.1068/a45311.

Wong, E., M. Jiang, Louise Munk Klint, Dale Dominey-Howes, and Terry DeLacy. "Evaluation of Policy Environment for Climate Change Adaptation in Tourism." *Tourism and Hospitality Research* 13, no. 4 (2013): 201–225. doi:10.2307/43575075.

World Bank. *The World Bank Group Action Plan on Climate Change Adaptation and Resilience*. Washington, D.C.: World Bank, 2019.

World Food Programme. *2018 - R4 Rural Resilience Initiative Annual Report*. 2018.

Zougmore, Robert, Samuel Partey, Mathieu Ouedraogomidele Omitoyin, Timothy Thomas, Augustine Ayantunde, Polly Ericksen, Mohammed Said, and Abdulai Jalloh. "Toward Climate-Smart Agriculture in West Africa: A Review of Climate Change Impacts, Adaptation Strategies, and Policy Developments for the Livestock, Fishery, and Crop Production Sectors." *Agriculture & Food Security; London* 5 (2016): 1–16. doi: 10.1186 /s40066-016-0075-3.

Zscheischler, Jakob, Seth Westra, Bart Hurk, Sonia Seneviratne, Philip Ward, Andy Pitman, Amir AghaKouchak, David Bresch, Michael Leonard, Thomas Wahl, and Xuebin Zhang. "Future Climate Risk from Compound Events." *Natural Climate Change* 8 (2018): 469–477. doi:10.1038 /s41558-018-0156-3.

INDEX

Page numbers in *italics* indicate figures.

Printed and bound by CPI Group (UK) Ltd, Croydon, CR0 4YY

12/11/2024

14591708-0002